大飞机出版工程

总主编 顾诵芬

航空经济学系列

民用飞机设计和运营经济性及成本指数

Design and Operational Economics for Civil Aircraft and Cost Index

陈迎春 主 编

宋文滨 李晓勇 副主编

上海交通大学出版社
SHANGHAI JIAO TONG UNIVERSITY PRESS

内容提要

本书汇集了航空公司运营经济性分析与飞机设计的优秀学术论文,内容主要涵盖航空公司运营经济性及其对飞机设计的影响问题。经济性贯穿商用飞机全寿命周期的所有环节,是反映商用飞机竞争力的重要指标之一。航空公司运营经济性对市场分析、机队规划、飞机设计等都具有重要的影响,开展有关经济性的研究和应用工作对于发展航空经济学这一新的交叉学科具有重要的意义。本书可供从事相关工作的技术和管理人员使用,也可供决策人员参考。

图书在版编目(CIP)数据

民用飞机设计和运营经济性及成本指数 / 陈迎春主编. —上海:
上海交通大学出版社,2013
(大飞机出版工程·航空经济学系列)
ISBN 978 - 7 - 313 - 10086 - 3

Ⅰ.①民… Ⅱ.①陈… Ⅲ.①民用飞机—设计 Ⅳ.①V271

中国版本图书馆 CIP 数据核字(2013)第 163496 号

民用飞机设计和运营经济性及成本指数

主　　编:陈迎春			
出版发行:上海交通大学出版社		地　　址:上海市番禺路 951 号	
邮政编码:200030		电　　话:021 - 64071208	
出 版 人:韩建民			
印　　制:浙江云广印业有限公司		经　　销:全国新华书店	
开　　本:787mm×1092mm　1/16		印　　张:12.5	
字　　数:248 千字			
版　　次:2013 年 11 月第 1 版		印　　次:2013 年 11 月第 1 次印刷	
书　　号:ISBN 978 - 7 - 313 - 10086 - 3/V			
定　　价:56.00 元			

大飞机出版工程

丛书编委会

总主编：

顾诵芬（中国航空工业集团公司科技委副主任、两院院士）

副总主编：

金壮龙（中国商用飞机有限责任公司董事长）

马德秀（上海交通大学党委书记、教授）

编　委：（按姓氏笔画排序）

王礼恒（中国航天科技集团公司科技委主任、院士）

王宗光（上海交通大学原党委书记、教授）

刘　洪（上海交通大学航空航天学院教授）

许金泉（上海交通大学船舶海洋与建筑工程学院工程力学系主任、教授）

杨育中（中国航空工业集团公司原副总经理、研究员）

吴光辉（中国商用飞机有限责任公司副总经理、总设计师、研究员）

汪　海（上海交通大学航空航天学院副院长、研究员）

沈元康（国家民航总局原副局长、研究员）

陈　刚（上海交通大学副校长、教授）

陈迎春（中国商用飞机有限责任公司常务副总设计师、研究员）

林忠钦（上海交通大学常务副校长、院士）

金兴明（上海市经济与信息化委副主任、研究员）

金德琨（中国航空工业集团公司科技委委员、研究员）

崔德刚（中国航空工业集团公司科技委委员、研究员）

敬忠良（上海交通大学航空航天学院常务副院长、教授）

傅　山（上海交通大学航空航天学院研究员）

总　序

　　国务院在 2007 年 2 月底批准了大型飞机研制重大科技专项正式立项,得到全国上下各方面的关注。"大型飞机"工程项目作为创新型国家的标志工程重新燃起我们国家和人民共同承载着"航空报国梦"的巨大热情。对于所有从事航空事业的工作者,这是历史赋予的使命和挑战。

　　1903 年 12 月 17 日,美国莱特兄弟制作的世界第一架有动力、可操纵、重于空气的载人飞行器试飞成功,标志着人类飞行的梦想变成了现实。飞机作为 20 世纪最重大的科技成果之一,是人类科技创新能力与工业化生产形式相结合的产物,也是现代科学技术的集大成者。军事和民生对飞机的需求促进了飞机迅速而不间断的发展,应用和体现了当代科学技术的最新成果;而航空领域的持续探索和不断创新,为诸多学科的发展和相关技术的突破提供了强劲动力。航空工业已经成为知识密集、技术密集、高附加值、低消耗的产业。

　　从大型飞机工程项目开始论证到确定为《国家中长期科学和技术发展规划纲要》的十六个重大专项之一,直至立项通过,不仅使全国上下重视起我国自主航空事业,而且使我们的人民、政府理解了我国航空事业半个世纪发展的艰辛和成绩。大型飞机重大专项正式立项和启动使我们的民用航空进入新纪元。经过 50 多年的风雨历程,当今中国的航空工业已经步入了科学、理性的发展轨道。大型客机项目其产业链长、辐射面宽、对国家综合实力带动性强,在国民经济发展和科学技术进步中发挥着重要作用,我国的航空工业迎来了新的发展机遇。

　　大型飞机的研制承载着中国几代航空人的梦想,在 2016 年造出与波音 B737 和空客 A320 改进型一样先进的"国产大飞机"已经成为每个航空人心中奋斗的目标。然而,大型飞机覆盖了机械、电子、材料、冶金、仪器仪表、化工等几乎所有工业门类,集成了数学、空气动力学、材料学、人机工程学、自动控制学等多种学科,是一个复杂的科技创新系统。为了迎接新形势下理论、技术和工程等方面的严峻挑战,迫切需要引入、借鉴国外的优秀出版物和数据资料,总结、巩固我们的经验和成果,编著一套以"大飞机"为主题的丛书,借以推动服务"大型飞机"作为推动服务整个航空科学的切入点,同时对于促进我国航空事业的发展和加快航空紧缺人才的培养,具有十分重要的现实意义和深远的历史意义。

2008 年 5 月,中国商用飞机有限公司成立之初,上海交通大学出版社就开始酝酿"大飞机出版工程",这是一项非常适合"大飞机"研制工作时宜的事业。新中国第一位飞机设计宗师——徐舜寿同志在领导我们研制中国第一架喷气式歼击教练机——歼教 1 时,亲自撰写了《飞机性能捷算法》,及时编译了第一部《英汉航空工程名词字典》,翻译出版了《飞机构造学》、《飞机强度学》,从理论上保证了我们飞机研制工作。我本人作为航空事业发展 50 年的见证人,欣然接受了上海交通大学出版社的邀请担任该丛书的主编,希望为我国的"大型飞机"研制发展出一份力。出版社同时也邀请了王礼恒院士、金德琨研究员、吴光辉总设计师、陈迎春副总设计师等航空领域专家撰写专著、精选书目,承担翻译、审校等工作,以确保这套"大飞机"丛书具有高品质和重大的社会价值,为我国的大飞机研制以及学科发展提供参考和智力支持。

编著这套丛书,一是总结整理 50 多年来航空科学技术的重要成果及宝贵经验;二是优化航空专业技术教材体系,为飞机设计技术人员培养提供一套系统、全面的教科书,满足人才培养对教材的迫切需求;三是为大飞机研制提供有力的技术保障;四是将许多专家、教授、学者广博的学识见解和丰富的实践经验总结继承下来,旨在从系统性、完整性和实用性角度出发,把丰富的实践经验进一步理论化、科学化,形成具有我国特色的"大飞机"理论与实践相结合的知识体系。

"大飞机"丛书主要涵盖了总体气动、航空发动机、结构强度、航电、制造等专业方向,知识领域覆盖我国国产大飞机的关键技术。图书类别分为译著、专著、教材、工具书等几个模块;其内容既包括领域内专家们最先进的理论方法和技术成果,也包括来自飞机设计第一线的理论和实践成果。如:2009 年出版的荷兰原福克飞机公司总师撰写的 *Aerodynamic Design of Transport Aircraft*(《运输类飞机的空气动力设计》),由美国堪萨斯大学 2008 年出版的 *Aircraft Propulsion*(《飞机推进》)等国外最新科技的结晶;国内《民用飞机总体设计》等总体阐述之作和《涡量动力学》、《民用飞机气动设计》等专业细分的著作;也有《民机设计 1000 问》、《英汉航空双向词典》等工具类图书。

　　该套图书得到国家出版基金资助,体现了国家对"大型飞机项目"以及"大飞机出版工程"这套丛书的高度重视。这套丛书承担着记载与弘扬科技成就、积累和传播科技知识的使命,凝结了国内外航空领域专业人士的智慧和成果,具有较强的系统性、完整性、实用性和技术前瞻性,既可作为实际工作指导用书,亦可作为相关专业人员的学习参考用书。期望这套丛书能够有益于航空领域里人才的培养,有益于航空工业的发展,有益于大飞机的成功研制。同时,希望能为大飞机工程吸引更多的读者来关心航空、支持航空和热爱航空,并投身于中国航空事业做出一点贡献。

2009 年 12 月 15 日

本书编委会

学术指导委员会
（按姓氏笔画排序）

王子方　叶叶沛　陈迎春
刘　洪　葛忠汉　黎先平

主　编

陈迎春

副主编

李晓勇　宋文滨

编委会
（按姓氏笔画排序）

杨　洋　宋文滨　李晓勇
余雄庆　陈晓和　俞金海

前　　言

飞机制造和民用航空是国民经济的重要产业,具有高技术、高附加值和知识密集性等特点,也是体现国家全球竞争力的行业。我国大型客机已经作为国家重大科技专项之一正式实施,在激烈的国际竞争条件下,我国发展民用飞机产业在关键技术、管理方式、适航取证、市场服务和人才培养等方面面临着长期的挑战。

民用飞机(民机)发展的历史就是通过持续性的关键技术与设计创新,不断提高其安全性、经济性、舒适性和环保性的过程。发展民机需要确保安全性达到适航标准,舒适性不断提高,对环境的影响逐渐减少。发展规律表明,每一代新机型的推出,需要在使用经济性上有 10%～15% 左右的改善,才可能获得航空公司认可。但是,飞机制造商在新技术研发上的投入,势必会推高飞机成本,在飞机价格主要由市场决定的情况下,进而降低其自身的盈利能力,影响其对后续机型的投入和持续稳健的发展。过分追求新技术甚至会导致飞机制造商在商业上面临破产的境地。对民用飞机经济性的研究涉及飞机设计、制造、使用、维护,以及处置等全寿命周期的各个环节,包含市场预测与分析、飞机定价、设计方案、供应商成本管理、航空公司竞争性分析以及新技术的经济性评估等广泛的内容,覆盖经济学、飞机设计、管理科学等多学科的综合交叉,形成航空经济学的主体。

航空经济学研究的内容应涵盖飞机设计和制造的经济性分析,民用航空经济性评估方法、标准和体系;飞机全寿命成本、直接使用成本、直接维修成本;影响飞机经济性设计的因素分析;面向价值工程的设计方法与体系;新一代布局飞机经济性分析;环保与经济性的关系;市场销售策略;经济环境和运营环境的影响,国家宏观政策和发展战略以及国际环境的影响等。

2012 年 12 月中国商飞市场研究中心联合上海交通大学在上海召开了第三

届"民用飞机经济性设计学术会议",本次会议的主要议题包括:

(1) 民用飞机经济性分析及设计研究;

(2) 民用飞机运营经济性分析;

(3) 民用飞机经济性评估方法、标准与体系;

(4) 影响飞机经济性设计的因素分析;

(5) 国家产业政策和经济环境对民用飞机产业的影响;

(6) 环保性对民机经济性设计的影响;

(7) 新技术对民用飞机设计的影响。

会议代表来自西飞国际、中国商飞、东航货运、南京航空航天大学,以及上海财经大学等公司和学校。针对民机经济性设计的相关议题,与会代表介绍了各自的研究成果,展开了热烈的讨论,本书即为本次会议的优秀论文成集而来。

本书可供从事相关工作的技术和管理人员使用,也可供决策人员参考。

目　　录

航空装备修复性维修费用的系统动力学仿真

刘 锦[1] 邵光兴[2] 黄兆东[3]

（1. 中国航空工业经济技术研究院财务管理部，北京，100012）
（2. 中国商飞上海飞机设计研究院，上海，201210）
（3. 中国航空工业发展研究中心财经研究部，北京，100012）

摘要： 本文首先分析了维修保障费用产生的因果关系；然后绘制存量流量图，建立存量、流量、辅助变量、常量之间的关联方程以及确定它们的初始值；在确定模型合理的基础上，分析模型的计算结果，并通过调整模型的参数来观察模型结果的变化以确定最优策略，为决策者提供决策支持。着重分析了飞机的零部件故障率、预防性维修的频率等对寿命周期内装备的维修保障费用的影响。分析结果表明本仿真方法具有一定的实用性。

关键词： 航空装备；修复性维修；维修费用；系统动力学；仿真

A research on the simulation approach based on system dynamics for aviation equipment's cost of repairing maintenance

Liu Jin[1]   Shao Guangxing[2]   Huang Zhaodong[3]

（1. AVIC Academy of Economics Technology Research，Beijing，100012）
（2. COMAC Shanghai Aircraft Design and Research Institute，Beijing，201210）
（3. AVIC Development Research Center，Beijing，100012）

Abstract： This paper first analyzes the causality of the maintenance and support (M & S) cost's generation; then draws the stock and flow diagram, establishes the correlation equation of stock, flow, instrumental variables, constants and determines their initial value; on the basis of the model reasonable, analyzes the model's calculation results, observes the change of the model results by adjusting model's parameters to determine the optimal strategy, provides decision support for decision-makers. Emphatically analyzes the influence of parts' failure rate and preventive maintenance rate on maintenance costs during the equipment's life circle. The analysis results have verified this paper's simulation method's practicality.

Key words： Aviation Equipment; Repairing Maintenance Engineering; Maintenance Cost; System Dynamics; Simulation

新时期航空装备的研制费和维修费都是高昂的,美军每年的装备维修费用高达200多亿美元,20世纪80年代以来美军航空装备维修费接近其研制费和采购费之和,其装备维修费约占国防费用的14.2%[1]。因此研究维修保障费用发生和发展变化的规律,研究降低和控制维修保障费用的有效途径,是控制寿命周期费用的重要手段。

现有维修保障费用的分析,往往考察长时域内维修保障系统实施的总体平均,费用结构相对固定,且为简化分析常常忽略维修要素之间的联系[2]。如果能从维修保障要素之间联系的定量刻画出发,通过分析维修过程中要素之间的相互作用和约束,能够全面描述装备维修过程及其暂态行为,那就可以对维修保障费用的产生进行解释,从而从费用的角度为维修保障系统的分析、设计和改进指明方向[3]。

本研究就是以降低费用为目标,以航空装备修复性维修过程为主要研究对象,应用系统动力学原理分析航空装备维修费用产生的动力学特性,建立起描述装备维修保障过程的通用模型,并运用计算机进行仿真分析,为装备维修保障费用的预测、分析和控制提供有效手段。

1　飞机修复性维修及其费用影响因素

1.1　飞机的修复性维修

修复性维修也称修理或排除故障维修。它是装备(或其部分)发生故障或遭到损坏后,使其恢复到规定技术状态所进行的维修活动。它可以包括下述一个或全部活动:故障定位、故障隔离、分解、更换、再装、调教、检验以及修复损伤件等。由此我们可以看出,影响装备修复性维修的因素有多方面。首先是装备的可靠性——它决定了在一定的时间内装备的故障次数,这与修复性维修费用有直接关系。其次是装备的测试性——它涉及能在多长时间内检测、定位到故障,这直接影响到人工时。另一个因素是装备的维修性。当装备或其部分维修性好时,维修人员能够在较短的时间内完成对故障的修复,也可以动用较少的工具,同时因为维修性好或许能节省部分材料。因此在本节考虑修复性维修保障活动时需要考虑如上所述的这些因素。

另外,影响飞机故障发生的因素还有飞机的任务强度与使用环境。在一定时间内飞机需要执行的起飞次数增多,飞行时间较长时飞机的故障必然会增加。当飞机的使用环境较为严酷,如需要面对低温、高温、高海拔、高湿度等环境问题时,也会影响到飞机的故障频率。因此在估算飞机的修复性维修费用时也应考虑这些因素。

1.2　修复性维修费用的影响因素

因为各类维修活动对应消耗相应的资源,各类维修活动之间可能有较大差别。因此,要做到对维修费用的精确估算,首先就需要了解具体型号的维修保障活动,这些信息在装备的维修方案中都能找到。当对新研装备应用系统动力学方法进行寿命周期维修保障费用评估时,可参考类似装备的维修保障活动进行建模。

飞机的本身组成非常复杂,有数万个零组件。不可能对所有的组成,所有的活动都进行仿真,这也是不可能完成的任务,事实上系统动力学方法也不建议对系统所有的细节进行建模。建模者可通过与专家、维修保障工人、维修活动记录人员、财务人员等进行交流,确定几类主要的形式,只对这几类进行建模,其他产生费用的维修活动可以简化或者折算到这几类中,仅当有必要细分时才考虑更进一步的仿真。

　　图 1 是本节针对修复性维修问题所描述的一个简单的例子。假设修复性维修费用是由可修有寿命件修复性费用、可修无寿命件修复性费用、不可修件费用组成。各部分组成及其详细的影响因素如图 1 所示。在 Vensim 软件中绘制成因果关系图如图 2 所示。

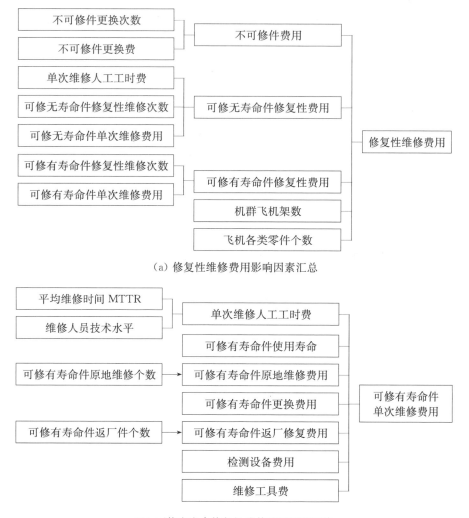

（a）修复性维修费用影响因素汇总

（b）可修有寿命修复性维修费用影响因素

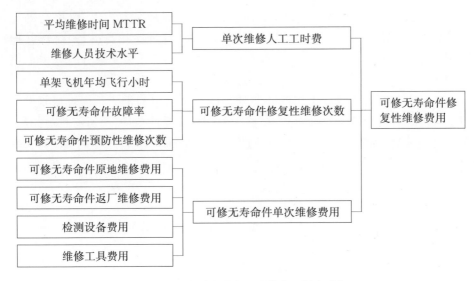

（c）可修无寿命件修复性维修费用影响因素

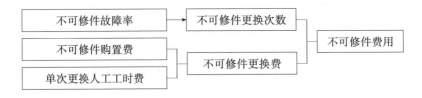

（d）不可修件修复性维修费用影响因素分析

图 1　修复性维修影响因素

2　修复性维修费用的因果关系分析

因果关系分析对建立准确、合理的系统动力学模型至关重要。因果回路图（causal loop diagram，CLD）是表示系统反馈结构的重要工具[4]。CLD 可以迅速表达关于系统动态形成原因的假说，引出并表达个体或团队的心智模型。如果认为某个重要反馈是问题形成的原因，就可以用 CLD 将这个反馈传达给他人。

一张因果回路图包含多个变量，变量之间由标出因果关系的箭头所连接。在因果回路图中也会标出重要的反馈回路。图 1 中列出的修复性维修费用影响因素的因果关系回路如图 2 所示。

3　修复性维修费用产生过程的存量流量分析

在社会、经济和生态系统中，存量和流量是两种最基本的变量。存量是积累，表明系统的状态并为决策和行动提供信息基础。流量则反映了存量的时间变化，流入

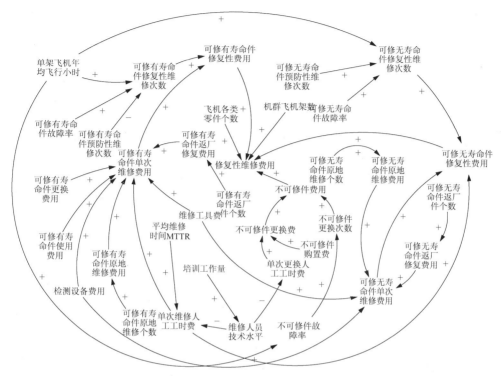

图 2　修复性维修费用产生的因果关系图

和流出之间的差异随着时间累积而产生存量。

　　由于我们研究的装备维修保障系统非常复杂,影响维修保障费用的因素多种多样。因此,若要在一个图中就将维修保障系统费用产生过程的存量流量图表述清楚不是一件容易的事情,而本文采用的流率基本入树建模法能够较好地解决该问题[5,6,7]。

　　修复性维修费用产生过程的存量流量图中的存量包括可修有寿命件修复性维修次数、可修有寿命件费用、可修无寿命件修复性维修次数、可修无寿命件费用、不可修件更换次数、不可修件更换费用和修复性维修费用。流量包括可修有寿命件修复性维修次数变化量、可修有寿命件费用变化量、可修无寿命件修复性维修次数变化量、可修无寿命件费用变化量、不可修件更换次数变化量、不可修件更换费用变化量和修复性维修费用变化量。另外还有可修有寿命件单次修复费用、单次维修人工工时费、可修无寿命件单次维修费用、不可修件单次更换费用等辅助变量,可修有寿命备件更换费用、可修有寿命件使用寿命、平均修复时间 MTTR,可修有寿命件现场修复率等常量。各量之间的关系及详细列表如图 3、图 4、图 5、图 6 和表 1 所示。

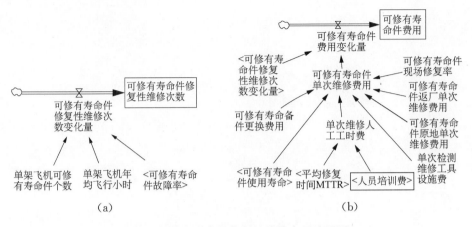

图 3 可修有寿命件修复性维修存量流量图

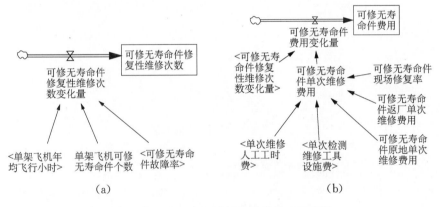

图 4 可修无寿命件修复性维修存量流量图

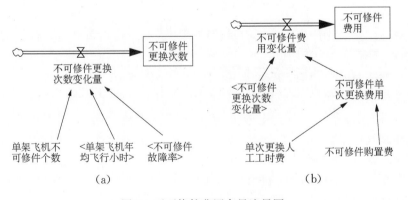

图 5 不可修件费用存量流量图

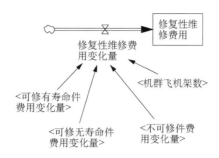

图 6 修复性维修费用存量流量图

表 1 修复性维修费用产生过程的关联方程

关联方程	水平变量方程(L)	(1) 可修有寿命件修复性维修次数＝INTEG(可修有寿命件修复性维修次数变化量) (2) 可修有寿命件修复性维修费用＝INTEG(可修有寿命件修复性维修费用变化量) (3) 可修无寿命件修复性维修次数＝INTEG(可修有寿命件修复性维修次数变化量) (4) 可修无寿命件修复性维修费用＝INTEG(可修无寿命件修复性维修费用变化量) (5) 不可修件修复性维修次数＝INTEG(不可修件修复性维修次数变化量) (6) 不可修件修复性维修费用＝INTEG(不可修件修复性维修费用变化量) (7) 修复性维修费用＝INTEG(可修有寿命件修复性维修费用变化量＋可修无寿命件修复性维修费用变化量＋不可修件修复性维修费用变化量)
	速率变量方程(R)	(1) 可修有寿命件预防性维修次数变化量＝单架飞机可修有寿命件个数＊单架飞机年均飞行小时/可修有寿命件故障率 (2) 可修有寿命件修复性维修费用变化量＝可修有寿命件修复性维修次数变化量＊可修有寿命件单次维修费用 (3) 可修无寿命件修复性维修次数变化量＝单架飞机年均飞行小时＊单架飞机可修无寿命件个数/可修无寿命件故障率 (4) 可修无寿命件维修费用变化量＝可修无寿命件修复性维修次数变化量＊可修无寿命件单次维修费用 (5) 不可修件更换次数变化量＝单架飞机不可修件个数/不可修件故障率＊单架飞机年均飞行小时 (6) 不可修件费用变化量＝不可修件单次更换费用＊不可修件更换次数变化量 (7) 修复性维修费用变化量＝(不可修件费用变化量＋可修无寿命件费用变化量＋可修有寿命件费用变化量)＊机群飞机架数

关联 方程	辅助变量 方程(A)	(1) 可修有寿命件单次维修费用＝单次检测维修工具设施费＋单次维修人工工时费＋(可修有寿命现场修复率 * 可修有寿命件原地单次维修费用＋(1－可修有寿命件现场修复率 * (可修有寿命备件更换费用＋可修有寿命件返厂单次维修费用)))* 可修有寿命件购置费/可修有寿命件使用寿命 (2) 单次维修人工工时费＝100 * 平均修复时间 MTTR * 人员培训费^LN (0.2) (3) 可修无寿命件单次维修费用＝单次维修人工工时费＋单次检测维修工具设施费＋可修无寿命件现场修复率 * 可修无寿命件原地单次维修费用＋(1－可修无寿命件现场修复率) * 可修无寿命件返厂单次维修费用 (4) 不可修件单次更换费用＝不可修件购置费＋单次更换人工工时费 (5) 可修有寿命件瞬时可用度＝1/可修有寿命件瞬时故障率/(平均修复时间 MTTR＋1/可修有寿命件瞬时故障率)
	常量方程 (C)	(1) 可修有寿命件现场修复率＝0.4 (2) 可修有寿命件原地单次维修费用＝600 (3) 可修有寿命件返厂单次维修费用＝800 (4) 可修有寿命件单次更换费用＝100 (5) 单次检测维修工具设施费＝80 (6) 平均修复时间 MTTR＝0.6 h (7) 单架飞机年飞行小时＝300 h (8) 机群飞机架数＝24 (9) 不可修件故障率＝[(0, 0)－(30, 10)], (0, 3e−005), (6, 3e−005), (12, 9e−005), (20, 0.000 27), (20, 0.000 27) (10) 可修无寿命件故障率＝[(0, 0)－(30, 10)], (0, 3e−005), (6, 3e−005), (12, 9e−005), (20, 0.000 27), (20, 0.000 27) (11) 可修有寿命件故障率＝[(0, 0)－(30, 10)], (0, 3e−005), (6, 3e−005), (12, 9e−005), (20, 0.000 27), (20, 0.000 27) (12) 不可修件购置费＝200 (13) 可修无寿命件现场修复率＝0.4 (14) 可修无寿命件原地单次维修费用＝600 (15) 可修无寿命件返厂单次维修费用＝800 (16) 单次更换人工工时费＝60 (17) 可修有寿命件购置费＝200
	初始值方程(N)	(1) 可修有寿命件修复性维修次数＝0 (2) 可修有寿命件修复性维修费用＝0 (3) 可修无寿命件修复性维修次数＝0 (4) 可修无寿命件修复性维修费用＝0 (5) 不可修件修复性维修次数＝0 (6) 不可修件修复性维修费用＝0 (7) 修复性维修费用＝0

4 案例研究

本案例对飞机寿命周期维修保障费用进行系统动力学分析。关于该飞机的所有数据均为假设,但这个系统内的各类数据之间的关联关系是可信的。

事实上,对于一个机群来说,影响其维修保障费用的因素多种多样。因此建立一个庞大、复杂系统动力学仿真模型需要有丰富经验的建模者,同时还需要对装备的维修保障过程及其费用非常熟悉的合作者才能建立一个真正具有价值的模型。因为本课题的时间、经费有限,本文仅作了相关的基础性研究,提出一个解决该类问题的可行方法,所以采用假设的案例。

4.1 案例假设

所作的基本假设如下所示:

(1)本模型中发生的所有费用均认为是现值,不再考虑通货膨胀因素。

(2)机群有 24 架飞机,每架飞机服役期均为 20 年,每架飞机平均年飞行 300 小时。

(3)该机群的维修保障费用由预防性维修费用、修复性维修费用、备件存储管理费用和维修人员培训保障费用组成。

(4)预防性维修仅考虑定时维修且时间间隔随着飞机的老化会作一定的调整;修复性维修仅包括零部件的修复和更换;备件存储管理费用仅受备件数量影响,即每个备件的存储费用是定值;人员培训费用每年的投入费用固定。

(5)假定每架飞机有 100 000 个零件,并且仅有三类零件:可修有寿命件、可修无寿命件、不可修件。其中可修有寿命件 60 000 个,可修无寿命件 20 000 个,不可修件 20 000 个。

(6)所有零件的失效率都一样。

(7)可修有寿命件、可修无寿命件、不可修件三类的单次维修费用都是固定值,检测、维修的工具/设施费也为定值。

(8)飞机故障件现场维修率为固定值。

(9)当某个部件故障后,采用修复或更换的措施后其可靠性提高,视为元部件的可靠性提高,即某个部件被更换后仍视为原部件的延续,只是其瞬时故障率降低了。

其他还有一些细节性的假设,在存量流量关联方程中直接体现,不再一一阐述。

4.2 修复性维修费用子系统流量存量关联方程

本模型在对修复性维修费用子系统建模时将飞机的零件分为三类:可修有寿命件、可修无寿命件、不可修件。分别对这三类讨论其修复及更换的人力费、材料费、工具费、设施费、备件更换费等等。具体的关联方程及其初始值等信息在表 1 中详细列出。

4.3 仿真结果分析

以故障率对维修保障费用的影响为例介绍本研究的仿真结果。

故障率是指产品工作到某时刻尚未发生故障,在该时刻后单位时间内发生故障的概率。这是影响装备使用的关键因素之一,也是影响装备维修保障费用的关键因素之一。换句话说,之所以对装备进行维修(包括预防性维修和修复性维修)就是因为装备会发生故障。本节将会在飞机原有假定的故障率的基础上分别下调33.3%、66.7%和90%,以观察当故障率降低时对飞机的寿命周期维修保障总费用及预防性、修复性维修费用的影响,如图7、图8所示。

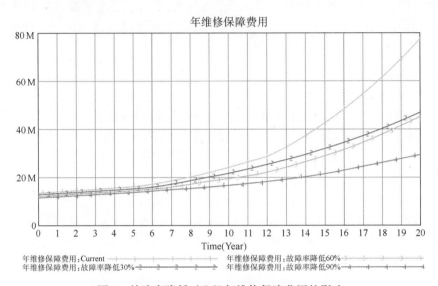

图 7　故障率降低对飞机年维修保障费用的影响

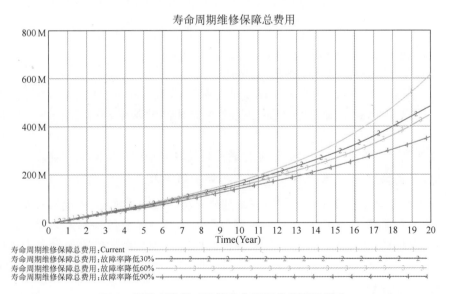

图 8　故障率降低对飞机寿命周期总费用的影响

当飞机的各零部件的故障率降低时其维修保障费用在中后期(图 7 中的 10 年之后)相对于较高的故障率时的装备维修保障费用明显较低,且增长缓慢。因此应尽量降低飞机的各组成零部件的故障率,提高其可靠性,以此来获得较低的飞机寿命周期维修保障费用。当然,降低飞机零部件的故障率必将提高飞机的设计和制造成本,这是需要在飞机设计阶段所权衡的。本文只是研究飞机的维修保障费用部分,因此通过系统动力学模型得到的结论是尽量降低故障率、提高可靠性以获取较低的维修保障费用。

5　结论

通过本文基于假设数据的案例研究,着重分析了典型装备(飞机)的零部件故障率对寿命周期内装备的维修保障费用的影响,认为装备零部件的故障率对其维修保障费用的影响较大,通过降低装备零部件的故障率能够有效地降低装备寿命周期费用。

由于时间较短,本课题在对装备维修保障构成与要素分析及应用系统动力学方法建模上仍有一些不足。首先是本文中建立的系统动力学模型针对的是一个简化的维修保障系统,现实中诸多复杂、数量庞大的因素都被忽略掉,而很可能恰恰是这些被忽略的因素是影响装备寿命周期内维修保障费用的关键因素,这样我们建立的模型就没有参考的意义了。当然,本课题是基础性理论研究,提供给读者的是分析该类问题的思路和方法,这也是本课题的意义所在。其次,本文所建立的模型是基于假设的数据。而对于系统动力学方法来说,初始值及相关变量的变化规律可能导致系统在仿真期间发生剧烈的变化,即准确的初始值及相关变量的已知变化规律是非常重要的。最后需要说明的是本模型的研究对象是装备的维修保障系统,而维修保障系统又是装备全寿命周期(装备的研制与生产、使用与保障、退役处理)范围内的一个子系统,对本子系统来说是最优策略的决策,但在全系统中可能并非最优(例如,本模型认为飞机的零部件故障率越低其维修保障费用越小,所以应尽量降低其故障率,但相应却增加了飞机前期的研制与生产成本)。因此利用系统动力学方法解决问题时一定要认清我们所要解决问题的边界。

综上所述,虽然具有部分局限性,但本文提供了一套应用系统动力学方法解决影响因素复杂、持续时间长的装备维修保障系统费用估算的有效方法,也对拓展装备寿命周期费用的评估手段提供了有益的帮助。

致谢:本文属于财政部重点课题"国产大型商用飞机全寿命周期成本控制模型与体系研究"(课题编号:2011QY013)和航空基金(基金编号:2011ZG28001)的阶段成果,在此对财政部课题基金和航空基金表示深深感谢。

本文课题的研究过程中,得到了北京航空航天大学常文兵老师的大力指导,在此表示深深感谢。

参 考 文 献

[1]　李郑琦，陈跃良. 飞机视情维修策略及其模型研究[J]. 航空科学技术，2011(3)：28 - 30.

[2]　Zhang Wenjin，Kang Rui，Guo Linhan，et al. Study on Military Equipment Support Modeling and Simulation [J]. Chinese Journal of Aeronautics，No. 2，May. 2005，18：142 - 146.

[3]　Christopher D Purvis. Estimating C - 17 Operating and Support Costs：Development of a System Dynamics Model [R]. Air force institute of technology，2001.

[4]　Narciso F Macia，George J Thaler. 动态系统建模与控制[M]. 李乃文，孙江宏，译. 北京：清华大学出版社，2006.

[5]　贾仁安，伍福明，徐南孙. SD 流率基本入树建模法[J]. 系统工程理论与实践，1998，18(6)：90 - 95.

[6]　贾仁安，胡玲，丁荣华，等. 系统动力学简化流率基本入树模型及应用[J]. 系统工程理论与实践，2001，21(10)：137 - 144.

[7]　胡玲，贾仁安. 强简化流率基本入树模型及枝向量矩阵反馈环分析法[J]. 系统工程理论与实践，2001，21(11)：83 - 88.

全虚饰服务航空公司与低成本航空公司运营模式对比和发展趋势分析

姜千祎[1]

（1. 苏格兰圣安德鲁斯大学）

摘要：这是一份经过研究全虚饰服务航空公司和廉价航空公司两种运营模式后对它们的现状进行分析和对比，并对未来发展进行展望的报告。全虚饰服务航空公司与廉价航空公司的运营模式的不同在于它们各自所采用的中枢辐射式航线网络模式和点对点航线模式的不同。前者能通过庞大的网络所产生的巨大经济效应获利并形成垄断趋势，而后者则在许多方面降低了成本，从而提供更便捷的服务，获得了旅客的青睐。虽然这两种运营模式之间存在竞争关系，但是从长期的角度看，未来的发展趋势应该是两种航空公司各自经营自己的市场，通过合作获得最大利益。

关键词：低成本航空公司；全虚饰服务航空公司；中枢辐射；点对点

Operational Comparison and Future Trend for Full-Service and Low Cost Airlines

Jiang Qianyi[1]

（1. University of St Andrews，Scotland，UK）

Abstract：A research into the analysis and comparisons between full-service and low cost airlines are given，along with some overview for future trend. The differences in network structure show that hub-and-spoke networks are typical for major airlines，while low-cost airlines often use point-to-point network structure. The operational patterns for the latter reduce cost，provide more frequent services and therefore attracted more passengers. It is predicted that two types of operations will continue to develop and maximum profit maybe achieved with a hybrid of the two.

Key words：The Low Cost Carriers；The Full Service Airline；The Hub-and-spoke Distribution Paradigm；The Point-to-point Paradigm

1 介绍

经营民航客运的航空公司（非货运）通常被分为全虚饰服务航空公司、低成本航空公司、包机航空公司以及支线航空公司4种。这4种航空公司代表了4种不同的运营模式。但是，今天的民航运输市场上包机航空公司以及支线航空公司的运营模式基本已趋同于前两种航空公司，尤其是低成本航空公司的模式。全虚饰服务航空及低成本航空通过市场的优胜劣汰过程而得到了巩固和发展，成为了主导市场的两大运营模式。

2　定义与比较

2.1　全虚饰服务航空公司

全虚饰服务航空公司得名于它们提供的种类繁多的飞行前及旅途中的虚饰服务，包括舱位选择，航班连接换乘等。此类航空公司多采用中枢辐射航线网络的运营模式，所以通常又被称为"中枢辐射式"航空公司。

2.2　低成本航空公司

低成本航空公司有时又称为廉价航空，得名于其提供的低廉票价。这类航空公司将运营的重点放在降低成本上，不提供虚饰服务。低成本航空公司所采用的运营模式被称为点对点模式，这种模式十分有利于节约各类成本，尤其是网络模式所带来的复杂性成本。

2.3　全虚饰服务航空公司与低成本航空公司运营模式对比

两种航空公司运营模式如表1所示。

表1　不同类型航空公司运营模式对比[①]

	全虚饰服务航空公司	低成本航空公司
机队组成	不同型号客机，种类较多	较为单一的飞机型号
航线网络覆盖	覆盖广泛，跨洲际网络	分散的国内支线或国际短程航线
航线网络结构	中枢辐射式航线网络，控制枢纽机场	点对点航线，使用次级机场
服务范围	全虚饰服务	无虚饰服务
定价	复杂性，歧视性定价	票价低廉，不包括托运行李等额外服务的费用
联盟	是全球航空联盟成员[②]	主要的低成本航空公司不是航空联盟成员

低成本航空公司模式与传统全虚饰航空公司模式最根本的不同是中枢辐射航线网络模式与点对点航线运营模式的不同。中枢辐射航线网络的运营理念是把相

[①] 全虚饰服务航空公司的前5个方面特点参照：Johannes Reichmuth，HansjochenEhmer，Peter Berster，Gregor Bischoff，Wolfgang Grimme，Erik Grunewald，Sven Maertens，"Analyses of the European Air Transport Market"，"Topical Report：Airline Business Models" German Aerospace Centre；Release 1.01，December 2008，p. 7.

[②] 沃纳·德夫曼、斯蒂芬·奥尔巴克、赫伯特·鲍姆与萨沙·阿尔伯斯主编，于剑等译"航空公司战略管理"，中国民航出版社，2008，中第一部分，2：托马斯·比格（Thomas Bieger）和桑德罗·安格斯特（Sandro Agosti），"航空公司业中的商业模式——发展与展望"p. 41

对稀薄的运输量集中到枢纽机场,而后在枢纽点最大化连接率,以及最大化可能连接的城市数量。大型航空公司的成功标志之一是拥有覆盖范围遍及全球的大型航线网络。这种模式为航空公司带来了各方面经济效应的优势,包括规模经济、密度经济和范围经济[①]。

联盟战略也是网络型航空公司的重要特点。欧洲的网络航空公司以"多枢纽航线网络"的运营为特征[②]。这是一种中枢辐射网络与联盟战略相结合的模式。由于枢纽分布密集,欧洲航空业的枢纽优势不如美国明显,因此航空公司之间采用结盟的战略模式,通过资源节省,扩大市场份额等协同增效作用来弥补各个枢纽间竞争所导致的经济损失。

网络型航空公司的运营模式主要的劣势在于其复杂性所带来的管理难度和复杂性成本。高峰时密集的航班流量对枢纽机场跑道容量及候机楼布局有很高的要求,还有包括行李处理系统等重要的基础设施的运行能力也受到极大考验。此外,高密度飞行导致机队、机组及地面服务人员利用率较低,机组培训及维护成本相对更高。这些成本被称为"基础相关复杂性成本"[③]。另一个主要成本则是"过程相关复杂性成本"[④],其中最重要的部分是航班延误时对飞机、旅客、行李的处理。

在联盟战略中,复杂性也是导致合作效率低下的重要原因。除了联盟内部的支线航班增加了枢纽运营的复杂性,联盟的运作还需要不同公司的规划人员参与复杂的决策过程来协调航班时刻。另外,联盟间许多合作如共享航班信息和共同的常旅客计划需要不同信息技术平台的系统整合,这也增加了复杂性。而且,根据现有的航空市场开放程度,航空公司之间的联盟通常会产生复杂的企业结构[⑤]。

与全虚饰服务航空公司的运营模式相对,低成本航空公司运营模式的核心特点是完全放弃对网络效应的利用,而采取分散的点对点航线运营模式。点对点的航线模式最大的特点是简化运营流程,并因此防止了复杂性成本的产生。在航线网络方面,由于无需枢纽连接,低成本航空公司可以避开航班密集的枢纽机场,而选择一些较偏僻,也相对便宜的小机场。又因为不提供航班连接服务,所以也免除了中转换乘所需的地面开支。

无虚饰服务的理念也是廉价航空的典型特征和减少成本的重要方面。最重要的是密集的座位使得每个航班上各成本门类的单位成本下降,航班利用率提升,增

---

① 沃纳·德夫曼、斯蒂芬·奥尔巴克、赫伯特·鲍姆与萨沙·阿尔伯斯主编,于剑等译"航空公司战略管理",中国民航出版社,2008,中第一部分,3;斯蒂芬·奥尔巴克(Stefan Auerbach),沃纳·德夫曼(Werner Delfmann),"欧洲航空公司业中网络型航空公司商业模式的巩固"p. 63 - 65

② 同上,p. 66

③ 沃纳·德夫曼、斯蒂芬·奥尔巴克、赫伯特·鲍姆与萨沙·阿尔伯斯主编,于剑等译"航空公司战略管理",中国民航出版社,2008,中第一部分,3;斯蒂芬·奥尔巴克(Stefan Auerbach),沃纳·德夫曼(Werner Delfmann),"欧洲航空公司业中网络型航空公司商业模式的巩固"pp. 67 - 68

④ 同上,pp. 67 - 68

⑤ 见参考文献 2, p. 29

加了经济性。在营销方面,廉价航空吸引大量价格敏感型消费者。机票只在网上直接销售,且不可退换,而且对信用卡消费和托运行李额外收费。

在节约资源,减少开支的同时,这种运营模式也减少了运营中发生错误的机率,从而提升了旅客的旅行质量。美国 2012 年一项关于旅客满意度的调查显示,在囊括了各航空公司登机、客机、机组、服务和飞行体验几项指标的排名中,低成本航空公司组以 754 分(满分 1000)领先于传统型航空公司的 647 分平均分,其中 JetBlue 以 776 分位居第一[①]。

3　两种运营模式的航空公司在飞机选型方面的特点比较

3.1　全虚饰服务航空公司

根据全虚饰服务航空公司的运营模式,它们在飞机选型方面的要求主要可以归结为以下 3 个方面:

1)机队的多样化

网络型航空公司需要不同类型的飞机来执行不同航程,以及满足不同服务的要求。而且,庞大的机队使用由不同飞机制造商所生产的飞机,以避免某一制造商对公司影响力过大。然而可以想象,在这样的情况下如果能适当加强飞机的通用性,则更受欢迎,因为这样有助于航空公司节约很大一部分由复杂机队所带来的成本。

2)飞机内饰

由于全虚饰服务和提供差异化服务战略的需要,航空公司对飞机内饰及座椅的要求将会越来越高,以体现自身品牌的个性化优势。

3)最新的科技

首先,由于受到低成本航空公司成本优势方面的竞争,全虚饰服务航空公司也需要在各方面降低成本来维持其竞争力,因此提高燃油效率将会成为最受关注的焦点之一。

其次,一方面由于差异化服务战略的需要,另一方面也由于国际上各类环境保护条约的出现,通讯、信息技术和环保领域上的技术革新,在飞机选型上将会越来越受到重视。尤其是夺人眼球的前沿创新科技更能增加差异化战略所需的卖点。

3.2　低成本航空公司

低成本航空公司在飞机选型方面主要有如下特征:

1)单一机队

与网络型航空公司相对,低成本航空公司因为航线单一,所以追求机队的机型

---

① Charles J. Johnson, "Be cheap: Low-cost carriers beat traditional airlines in customer satisfaction" Chicago Tribune Newspapers, June 14, 2012, http://articles. chicagotribune. com/2012-06-14/travel/ct-taking-off-lowcost-carriers-outpace-traditional-airlines-in-customer-satisfaction-20120613 _ 1 _ stuart-greif-low-cost-carriers-satisfaction

单一。航空公司选择购买同一生产商同一型号的飞机或选择通用性较高的飞机,由此便节约了机队组成多样所带来的复杂性成本,当一次性购买多架飞机时还能获得折扣。单一的机队组成也有助于节约维护和人员培训的成本。

　　2)高度的可靠性

低成本航空公司的客机使用率极高,需要在一天之内频繁地起降,由此带来了对飞机安全可靠性的极高要求,因此航空公司一般会选择较为成熟的飞机型号而不是着眼于最新的科技来作为卖点。

4　发展趋势展望

　　全球航空公司运营模式的发展主要可以归结为 3 种趋势:联盟化趋势、趋同化趋势和追求差异化服务的趋势。各国航空市场的持续放松管制以及越来越多双边飞行协议的取消,为国际航空联盟的扩大提供了条件。尤其是传统大型航空公司开始选择与小型支线航空公司结盟,通过非直接的方式来运营次级航线上的业务。在一些亚洲国家,航空市场的自由度提高也为低成本航空公司在越来越多国家的发展提供了条件。而在欧美等地,这些航空公司则开始出现逐渐向传统网络型航空公司运营模式转变的迹象。同时,大型的网络型航空公司则将重点放在提供差异化的服务上,来与其他网络型航空公司和低成本航空公司竞争。

4.1　联盟化趋势

　　联盟化是受市场自由度影响很大的一种航空公司战略。各国航空市场自由度的不断提高将会为跨国界和跨商业模式的航空公司联盟提供有利条件。正如之前所提到的,结盟是网络型航空公司最大化经济效应的一种有效,甚至有时是唯一的手段。而 2007 年至今的经济危机,以及 9/11 和非典型性肺炎等突发事件更使得航空公司意识到单独面对这些影响因素时自身的脆弱,因而迫使它们更加重视联盟战略。

4.1.1　跨国航空联盟

　　航空公司的跨国结盟在目前依然受到一些国家航空市场管制的限制。管制内容中重要的一项是各国互相签订只允许对方国家航空公司运行两国之间来往航线的双边协议。这样的协议曾经使得航空市场完全呈现分散化的态势。目前欧洲航空公司间整合程度最高的是法航和荷航,以及汉莎和瑞士航空。而它们合作所产生的大量额外成本,包括谈判成本和企业复杂性结构导致的成本使许多航空公司不敢尝试这样的整合模式。

　　跨越洲际的航空联盟则遇到更大的阻力。在许多国家,尤其在亚洲等地,航空市场的自由化程度并没有达到欧美的水平,因此给外国航空公司与这些国家的航空公司之间的联盟运营带来许多不利的因素。但需要指出的是,联盟战略并非在世界上所有地区都如在欧洲一样受到青睐。在一些情况下一个地区航空公司的战略也许受到地区文化不同的影响。欧洲在传统上倾向于一种一体化或结盟的运行模式,相信这会给国家和企业带来长期稳定的最大化的成功,加上欧盟一体化的关系,欧洲航空公司也许会更容易寻求深入的一体化模式。

然而,鉴于联盟战略给网络型航空公司带来的巨大的正面影响,人们依然有充分理由相信随着自由化的必然趋势,各地航空市场的环境会越来越适合航空联盟的发展,而采用联盟战略也会是各国网络型航空公司运营模式中的重要趋势。正是因为目前许多不具备的条件,使得将来联盟的发展存在着巨大的空间。

4.1.2　不同运营模式的航空公司之间的联盟

大型网络型航空公司与低成本航空公司或支线航空公司之间的兼并整合或结盟也是航空公司运营模式的发展趋势之一。然而在此过程中,低成本航空公司通常是被兼并或收购的对象。联盟战略理论上不会给点对点运营模式的航空公司带来经济效应和最优化的运营。目前世界上主要的低成本航空公司运捷和瑞安等都不是世界航空联盟的成员,更没有由廉价航空公司所组成的联盟。

相反,网络型航空公司的综合生产模式需要其在所有细分市场上拥有市场份额。然而特别在高需求的非枢纽航线上,全虚饰服务的成本复杂性与商务舱市场较低的需求使得网络型航空公司不能产生可持续利润基准。通过兼并低成本航空公司,或与之联盟,大型网络型航空公司便能够通过其低成本单元或联盟伙伴占有或分享那些它们没有竞争优势的非枢纽运输市场。比如 2006 年意大利的全虚饰服务航空公司 Alitalia 收购 Volare 航空[①]来对抗国内大量的外国廉价航空的竞争。而欧洲其他各大航空公司也都通过此类方式拥有自己的低成本单元。

网络型航空公司对支线或低成本航空公司进行的合并行为还出现在一些新开辟的航线上。比如在欧盟与美国签订开放天空协议不久,除法航外并没有许多网络航空公司立即开拓出直飞的航线。英国航空公司收购 L'Avion 来执行从伦敦及各大欧洲枢纽飞跃大西洋的航线。可以想象,由于这种战略对于网络型航空公司是十分有利的,因此只会被越来越多的航空公司所采纳。

4.2　差异化战略模式的趋势[②]

许多网络型航空公司都认为,为了应对低成本航空公司所带来的挑战,非常重要的一方面在于尽量发挥全虚饰服务带给旅客的旅程舒适度的优势,通过具有差异化的个性服务提供客户价值。这样的差异化战略一般从三个层面进行分区:产品销售;飞行时刻;座椅及航行中的虚饰性服务。

1) 产品销售
地面及客服与乘客的正面接触,行李,机舱服务。

---

① 见参考文献 2, p. 31 - 32
② 沃纳·德夫曼、斯蒂芬·奥尔巴克、赫伯特·鲍姆与萨沙·阿尔伯斯主编,于剑等译"航空公司战略管理",中国民航出版社,2008,中第一部分,4:金·夫勒斯科夫(Kim Flenskov)"从产品型到服务型——航空公司成功的差异化战略模式"p. 79 - 96

2）飞行时刻

较短飞行时间或飞行频率可使高票价合理化。

3）座椅及航行中的虚饰性服务

座椅舒适度、载重量等/机上娱乐系统及上网系统。

目前许多大型的航空公司都已开始提供几乎是同等水平的机上及地面服务，包括先进的机上娱乐设施、高舱位可以平躺的座椅、地面换乘等，同时立志于为乘客提供更加个性化和无间隙的服务。然而一些目标的实现，比如缩短飞行时间或更高的飞行频率等则需要航空公司克服自身运营模式本身的劣势，而达到在持续高强度的航班运营期间各部门流畅的配合和运作。

4.3 运营模式的趋同化趋势

除了一些网络型航空公司开始在新开通的航线市场上尝试点对点的运营模式，运营模式趋同化的更大趋势则是低成本航空公司开始向全虚饰服务网络型航空公司的模式转变。这种转变主要体现在三个方面：第一，某些低成本航空公司开始尝试全虚饰服务；第二，开始提供转机服务；第三，开始尝试开辟长途航线。

2007年早些时候，瑞安航空公司申明将在10年之内进入大西洋航线的市场，计划运营从其自己的欧洲基地到美国各个较小的机场之间的航班，并且将提供座舱服务和2个舱位的区分。事实上，一直以来，并非所有的低成本航空公司都完全遵循无虚饰服务的模式特点。比如之前提到过的美国 Jet Blue 航空公司就为自己提供的免费机上遥控电视，收音机和一个免费托运行李等虚饰服务而引以为傲。而在美国，由于所有的航空公司在国内航线上的托运行李都要额外收费，顾客体会到的网络型航空公司与低成本航空公司在虚饰服务方面的差异显得比其他地区要小得多，而 Jet Blue 也认为这些虚饰服务是成就他们极高的客户满意度的重要因素。

另一个更为重大的转变是某些低成本航空开始提供的转机换乘服务。传统上根据低成本航空公司点对点的运营模式，航空公司不会提供转乘连接的地面服务。然而搭乘廉价航空的相当一部分旅客确实存在换乘的需求，一些人甚至在机场搭起帐篷来等待第二天起飞的航班。美国西南航空公司在拉斯维加斯和芝加哥 Midway 机场的票务联通和行李直挂的转机服务已经持续了好几年，这两个机场庞大的客流量和高密度的航班流自然而然地为该公司运营这样的业务提供了机遇，不过西南航空公司并没有因此使得自己的运营最优化。

而多数业内人士不看好低成本航空公司的转机服务业务[①]。在欧洲一些航空公司如德国之翼所提供的转机服务中依然需要将行李取出后重新登记。而更重要的是，当两个支线航线之间的中转需求达到一定量时，市场自然会产生这两地的直飞航线，使得换乘变得没有意义，并且提供换乘服务使得点对点模式朝网络模式发展，自然而然带来一定复杂性成本，使得廉价航空运营成本上升，接近网络型航空公司，

---

① 见参考文献2，p. 36

这样就降低了它们最大的优势。

最后,一些人认为低成本运营模式也可以被运用到长距离航线上。不过这样的理论受到许多质疑。尽管曾经有一些廉价航空尝试开拓洲际长航线运营,但却几乎都以失败告终。就像第二部分中提到过的,长途航线使得航空公司各方面的运营成本提高,使得其无法再维持很低的票价,尤其是一些国家由于航空协议的限制,不允许在枢纽机场以外的机场开通国际航线,使得低成本航空无法使用次级机场。

由此可以再次证明,低成本航空公司的成功是建立在一定市场条件和运营模式下的,而脱离了这些市场条件或改变其运营模式便会使其丧失优势。将低成本模式转变为网络型模式的趋势是否有利于航空公司发展还有待观察。

5　中国航空公司的运营模式

目前中国的航空公司,尤其是全虚饰服务航空公司的运营模式较欧美的模式有很大的不同。造成这种不同的最根本原因是中国航空运输市场的政府管制力度高,自由化程度低。而且中国的三大骨干航空公司,即中国国际航空股份有限公司、中国东方航空和中国南方航空均属于国有企业,由国家来负担其亏损,所以导致公司及时调整战略来应对市场竞争的动力不足。

从全虚饰服务航空公司的角度来讲,管制力度大主要造成了中国大型航空公司在航线网络以及机队组成上与欧美航空公司不同。最明显的是中国航空市场的枢纽效应不明显。许多大型航空公司并没有像欧美的同行那样,因为增加经济效应的需要而发展中枢辐射航线网络,反而主要以运营点对点的航线为主。另外,由于人口众多而机场相对较少,许多航空公司共飞同一个城市对,共用同一个枢纽[①]。这样使得枢纽运营产生的经济效应大大降低,航线上航班过于拥挤,造成大量资源浪费。由于缺乏良好的竞争环境,航空公司未就这个问题做出战略上的调整,比如像欧洲的航空公司那样发展和完善多枢纽航线网络,以及联盟的战略模式。

航线运营上反映的另一个不符合自由市场的现象是一些航空公司被要求运营一些"政治航线",即由于国家考虑到外交以及支持一些地区经济发展的需要,而令航空公司开通一些无法盈利的航线[②]。

而从机队组成上来讲,飞机的购买和规划都由国家统一执行,一些决策可能出于政治原因,由此造成机型和运力与市场需求不完全符合,也造成与航空公司需求的不完全符合[③]。

大型航空公司的运营模式现状也直接影响到了中国低成本航空公司的发展。由于航空市场普遍运量过剩,并且受到国家支持的大型航空公司不惧怕降低票价来增加销售量和占领市场,低成本航空公司并没有便捷的直达航班和低价机票的优

---

① 杨思梁"航空公司的经营与管理"中国民航出版社,2007,pp. 26 – 27

② 同上,p. 29

③ 同上,p. 27

势。而最关键的是,由于政府的控制,航油航材等被垄断,导致航空公司刚性成本占比重较高,从而成本下降空间减少。这些原因导致中国的低成本航空公司的发展面临困难①。

6 总结

在欧美市场上,全虚饰服务航空公司和低成本航空公司是一种共生关系,拥有这两种运营模式的航空运输市场能够保持竞争活力,并成为一个健全的市场。在支线或短途航线上,低成本航空公司的模式更科学:效率更高,节约资源,减少航程中发生错误的因素。然而这种科学的模式是网络型航空公司由于其自身运营模式限制无法施行的。所以最好的方法是让廉价航空运营这部分市场,而网络型航空则专注于能体现它们优势的洲际航线,或干线航线市场。然而两者可以通过结盟等各种合作形式,互相分享对方在另一部分市场上的收益。这样,不仅双方都获得了最大的收益,也能使整个市场运行更加有效率,节约资源,提供更加优质的服务。

运营模式的发展趋势显示在将来很长一段时间里,全虚饰服务网络航空公司和廉价航空公司这两种模式将会继续存在,并能够被区分。但航空公司的运营模式并不再单一化。现有的趋势,包括联盟战略的持续发展,对差异化服务的重视都体现了网络型航空公司模式的巩固。但另一方面来讲,主要航空公司通过兼并或吸收加盟等方式吸收低成本航空的营运模式进入自己目前的主要运营模式中。也许这些整合的形式有助于保持这两类航空运营模式各自的鲜明特点:当大型的全虚饰服务航空吸收兼并了一些小型的廉价或支线航空公司,它们会令这些公司保持原有的经营模式来替它们占领那一部分的市场。而中枢辐射网络的那一部分将会持续贯彻传统的运营模式,占领高端市场。

值得一提的是,在其他一些新兴市场,如东亚航空市场,私有化程度仍有很大的上升空间。在欧洲,市场的放松管制产生航空公司运营模式的巨大改变,并逐渐形成现有的标准模式,而那些新兴市场的未来发展所带动的运营模式是否会有创新依然有待研究。

参 考 文 献

［1］ 沃纳·德夫曼,斯蒂芬·奥尔巴克,赫伯特·鲍姆与萨沙·阿尔伯斯主编,于剑,译.航空公司战略管理[M].北京:中国民航出版社,2008:41,63－65,66,67－68.
［2］ 杨思梁.航空公司的经营与管理[M].北京:中国民航出版社,2007:26－27,184.
［3］ 轩余恩.中国民航运输市场发展与创新[M].北京:中国民航出版社,2003.
［4］ Johannes Reichmuth, HansjochenEhmer, Peter Berster, Gregor Bischoff, Wolfgang Grimme, Erik Grunewald, Sven Maertens. Analyses of the European Air Transport

① 杨思梁"航空公司的经营与管理"中国民航出版社,2007, pp. 184

Market，Topical Report：Airline Business Models. German Aerospace Centre；Release 1. 01，December 2008：7，31 - 32，36，29，27.

[5] John F. O'Connell. The rise of the Arabian Gulf carriers：An insight into the businessmodel of Emirates Airline [J]. Journal of Air Transport Management 17（2011）339e346journal homepage：www. elsevier. com/locate/jairtraman.

[6] Charles J. Johnson. Be cheap：Low-cost carriers beat traditional airlines in customer satisfaction [J]. Chicago Tribune Newspapers，June 14，2012，http：//articles. chicagotribune. com/2012-06-14/travel/ct-taking-off-lowcost-carriers-outpace-traditional-airlines-in-customer-satisfaction-20120613_1_stuart-greif-low-cost-carriers-satisfactio.

飞机引进方式　综合评估策略

许　敏[1]　张　康[1]

(1. 上海飞机设计研究院财务部，上海，200232)

摘要：本文首先介绍飞机引进各种方式的特点，接着分析航空公司和租赁公司飞机引进考虑的各种因素：融资渠道，融资机构和飞机引进各种方式对所有权成本的影响，最后总结了航空公司飞机引进综合评估策略的特点。

关键词：融资租赁；经营租赁；资产负债率；现金流量；投资收益率

Assessment strategies for Civil Aircraft Acquisition Costing

Xu Min[1]　Zhang Kang[1]

(1. COMAC Shanghai Aircraft Design and Institute，Shanghai，200232)

Abstract：At first，this paper introduces kinds of aircraft procurement models，then analyzes airlines and leasing company decision-making reasons，financing mode，financing institution and procurement models influence to ownership cost. Finally the article summarizes aircraft procurement synthetical evaluation strategy in airlines.

Key words：Financing Lease；Operation Lease；Asset-debt Rate；Cash Flow；Rate of Investment Return

飞机作为大宗商品本身价格不菲，投入运营期限较长，产生收益和投入成本之间如何平衡？航空公司和租赁公司对飞机不同的持有目的又将对飞机引进产生何种影响？飞机销售过程中如何作好客户推介？飞机引进综合评估策略研究无疑是航空公司、租赁公司和制造商都必须深入研究的问题。

1　飞机引进方式特点分析

1.1　采购方式

采购方式下航空公司通常在飞机交付前两年就要预定，通常需要首付30％的预付款，待飞机交付时以飞机为抵押物获得银行贷款来支付给飞机制造商。这种方式特点是先期投入资金量大，向银行融资，银行贷款将增加公司债务并会导致资产负债率上升，航空公司作为直接运营方拥有飞机所有权，飞机作为固定资产计入航空公司财务报表，计提折旧年限较长，通常为20年，残值率5％，航空公司将承担飞机残值风险。

1.2　融资租赁

融资租赁飞机一般由租赁公司预定飞机,航空公司从租赁公司租飞机,因此交付期短,一般融资租赁年限在 8 年～12 年,每季度支付一次租金,当然租赁期开始还需要交纳维修保证金和押金。融资租赁飞机的特点是先前投入资金量小,向租赁公司融资,每期支付的是租金,租赁期满后航空公司才拥有飞机所有权,但是飞机可以先作为固定资产入账并每年计提折旧,折旧年限根据租期分摊,一般是 10 年,残值率30％。航空公司将承担飞机的残值风险。因此如果只考虑单一变量来看,不会导致航空公司的资产负债率上升。

1.3　经营租赁

经营租赁飞机通常是航空公司为满足热门航线的临时需要,因此租期较短。航空公司可以迅速的改变机队数量,经营租赁的主要特点是通过支付租金向租赁公司融物,租期较短一般为 3 年～5 年,租期内拥有使用权,作为表外资产不计提折旧,不承担飞机残值风险。

三种方式特点比较(见表 1)。

表 1　三种方式特点比较

采购方式	融资租赁	经营租赁
先期投入资金量大	分期付款方式采购飞机	支付租金
向银行融资	向租赁公司融资	向租赁公司融物
提高资产负债率	降低资产负债率	
拥有所有权	租期到后拥有所有权	拥有使用权
计提折旧	计提折旧	不计提折旧
承担飞机残值风险	承担飞机残值风险	不承担飞机残值风险

2　航空公司飞机引进考虑因素

2.1　资产负债率

资产负债率是衡量企业债务规模优良与否的关键性指标,对企业的融资能力和规模有着至关重要的影响。国内航空公司的资产负债率普遍偏高,是束缚航空公司以采购方式引进飞机的主要原因。国航、南航、东航等三大航空公司 2011 年的年报反映,期末资产负债率分别为 71％,71％,80％。

2.2　现金流量

现金流量是航空公司为保证日常经营需要必须留存的流动资金,在直接运营成本中,除了所有权成本以外的其他成本均为现金运营成本,包含空勤成本、维护成本、燃油成本、导航费、机场收费、餐食费、地面服务费、民航发展基金等。航空公司必须保持一定的现金流量来维持日常航线营运的需要。

航空公司预测现金流量首先预测运输量、上座率和客票收益水平,然后测算出满足市场需求的投入运力以及预计费用,从而测算出运营成本。现金支出预测包括资本性支出、购置飞机的分期付款、未来的股息和税金支出以及资产处置损益等等。在预测现金流时还需要包括相应的资金安排计划和利息支付计划。

2.3 税收减免政策

税收减免政策是航空公司飞机引进必须要考虑的因素之一,在融资租赁方式下,假设按租期10年计算,航空公司每年只付低于10%飞机价值的租金,却按全额飞机价值计算折旧,并且租金和折旧额都可以抵扣企业所得税。

航空公司对采购和融资租赁两种形式拥有飞机都要计提折旧,一般采购飞机的折旧期为20年,融资租赁飞机折旧期根据租赁期限来定,一般为10年,因此融资租赁飞机分摊到每年的折旧额比采购飞机多,抵扣所得税也相应多。

关税和增值税方面,不同国家飞机租赁的税务优惠政策将直接影响到租金的高低水平。境内租赁飞机,一般只有一家租赁公司享受减征关税和增值税的税务优惠,折旧抵税从而航空公司可以获得较低的融资成本;对跨境租赁飞机,很可能涉及转租赁,从而使得两家以上租赁公司享受税务优惠,只要符合所在国的税法规定同一架飞机可以在不同国家提取折旧,承租人获得较低的税后融资成本。

2.4 飞机的交付期

飞机交付期较长,飞机制造业对新飞机的交付期一般是3~4年,然而租赁飞机方式取得飞机一般只需要几个月,缩短了飞机的交付期。航空公司为满足季节性的临时需要往往首先要考虑飞机交付期。

2.5 机队构成

机队配置构成也是航空公司考虑的因素,航空公司通过开展飞机租赁业务,在航空客运量增长期,可以充分利用租赁杠杆手段,以适应拓展和开通航线的需要,解决航空运力的问题。在航空市场客运量萧条期经营租赁飞机灵活性较大,可以避免航空公司资产闲置,迅速改变航空公司的机队状况。

当然航空公司将结合公司运营状况和航线网络规划在机队配置中将这三种方式组合运用,以规避各种风险。

3 租赁公司飞机引进考虑因素

3.1 投资收益率

投资收益率是决定租赁公司从事此项业务的关键因素,租赁公司向银行融资采购飞机,租金的高低是由租赁公司的融资成本、税收政策和预期收益率决定的,融资成本高低决定了租金的高低,我国开展融资租赁业务通常是有银行业背景的租赁公司,优势在于可以获得较低的资金成本。国内的四大行都设立租赁公司,开展飞机租赁业务较强的有中银租赁、工银租赁、国银租赁、交银租赁和民生租赁。

3.2　飞机残值

租赁公司不直接运营飞机,飞机的残值将涉及二手飞机市场的买卖价格,二手飞机市场评估价值和飞机实际账面残值使得租赁公司可以从飞机残值处置中获得更多盈利空间。例如:B737飞机和A320飞机因为年产量的不同,导致二手机市场上B737飞机更具有保值。

3.3　市场需求

航空公司需求量大的热门飞机,通常是租赁公司乐于采购的。例如:B737和A320这样的热门飞机,年产量订单排满,机位无疑成了航空公司和租赁公司争夺的资源。

4　融资渠道

融资渠道是购置大宗商品必须考虑的重要环节,通常分为企业内部融资和企业外部融资。自筹资金率是从内部筹集资金占资本支出比例,当经济周期低谷时,航空公司现金流入少却面临大批飞机交付,这一比率很低,当现金流得到改善,投资较少时这一比率很高。

4.1　内部融资渠道

内部融资渠道可以分为股东投入和企业利润两方面,全球大部分航空公司仍由政府控制着50%以上的股份。股东有:政府,其他航空公司,金融机构,航空公司雇员,其他个人。企业利润通常由机票收益、货邮收益和其他业务收益构成。

4.2　企业外部融资渠道

企业外部融资渠道一般由以下五部分构成:银行贷款、应付账款、股票(债券)、制造商融资和租赁。

银行贷款是最常用的融资方式。

应付账款属于供应链融资的一部分,航空公司拖欠航油和机场收费,可以获得无息的资金来源。

在证券市场中上市的航空公司,通过市盈率就可以知道新的权益资本大概价格水平。市盈率高意味着新资本的成本低。当然发放公司债券也是一种融资手段。

国际上飞机制造商通常都会成立自己的金融机构,推动飞机销售工作。

通过租赁公司租赁飞机也是一种融资手段。

5　融资机构

参与融资活动的机构有银行、租赁公司和制造商。

5.1　银行

资金借贷双方中银行扮演着重要角色,银行贷给航空公司的资金来源有储户的存款和银行自有资金。银行资产负债表上会出现银行贷款,并会受到贷款限额和流动比率的限制。

大多数国家都设有自己的出口信贷机构,这些机构有的是政府的一个部门,有的是政府扶持的民间组织。这些机构通过担保或保险的形式(而不是直接放贷),支持本国商品的出口。当银行不愿意承担全额风险时,这些机构则为放贷提供支持或者补充。

以下出口信贷机构全部来自飞机制造生产大国,均参与航空公司的飞机融资活动(见表2)。

表 2　出口信贷机构名称

美国进出口银行(Ex-Im Bank)	日本出口与投资保险公司(NEXI)
英国出口信贷担保署(ECGD)	加拿大出口开发公司(EDC)
法国外贸保险公司(COFACE)	意大利出口信贷保险局(SACE)
德国赫尔梅斯信用保险集团(Euler Hermes)	

5.2　租赁公司

租赁公司通常分为两类:一是有银行业背景的租赁公司,优势在于可以获得较低的资金成本,从事融资租赁业务,如国内的四大行都设立租赁公司,开展飞机租赁业务较强的有中银租赁、工银租赁、国银租赁、交银租赁和民生租赁;二是飞机和发动机制造商背景的租赁公司,优势在于从事经营租赁和发动机租赁业务,从融物出发解决客户的融资问题,如GECAS金融租赁公司、波音金融租赁公司、空客金融租赁公司、庞巴迪金融租赁公司、巴西航空工业金融租赁公司。

5.3　租赁公司

飞机制造商可以通过赊销或回购飞机等方式向航空公司提供融资支持,有时也会采用贷款或股权投资方式。

波音公司1987年5月以债券形式向联合航空提供7亿美元贷款。

6　影响飞机所有权的相关因素

6.1　采购方式

飞机的价格是影响所有权成本的主要因素,商品价格按年分摊将影响折旧额的高低,汇率利率、税收和保险都和商品价格密切相关。

飞机等大宗商品涉及的贷款期限长,利率风险不可避免,实际银行贷款业务中由于企业多,资金供给少,各家商业银行经常按公布官方利率上浮15%～30%来放贷。

利息费用计算通常按每半年等额偿付借款本息。

汇率风险,是指经济主体在持有或运用外汇的经济活动中,因汇率变动而蒙受损失的可能性。飞机采购合同通常贷款年限较长(一般和折旧年限相同为20年),且合同大多以美元计价和结算,因此,汇率的变动将影响飞机的所有权成本。2004年我国实行"以市场供求为基础的、单一的、有管理的浮动汇率制"汇改以来,人民币形成机制弹性的增强,人民币汇率波动有所加大,其对美元的升值五年已达22%,年均超过40%。

折旧是对飞机的购机费用在其使用期间进行成本分配的一种方法,即对最初购机费用的分摊。

根据《企业会计准则第4号—固定资产》应用指南规定,应计提折旧额是指飞机账面原值减去预计净残值后的金额(已计提减值准备的,还应当扣除已计提减值准备累计金额。飞机在后续使用过程中发生更新改造支出的,被替换部分应该扣除其账面价值)。折旧方法:常用直线折旧法。

采购方式下折旧年限是根据固定资产的使用年限来确定的:飞机的一般使用寿命为20～30年,国内航空公司将折旧期订为20年。残值率:内资航空公司残值率5%,外资航空公司残值率10%。

飞机进口和航材备件进口涉及的税金主要有关税和增值税。航空公司按飞机的采购价格全额征税。

6.2　融资租赁

融资租赁根据美国会计准则的定义,在租赁过程中飞机的所有权转移给承租人符合以下三个条件之一的就可以算融资租赁:①租期占到资产经济寿命的75%以上;②最低租赁付款额的现值占到飞机公允价值的90%以上;③承租人拥有廉价购买选择权(BPO)。

融资租赁租金的构成:设备价款＋融资成本＋营业费用＋利润。

保证金支付数量和方式:相当于3个月的租金的履约保证金。

出租人购买航空公司选定机型,享有所有权,在一定期限内将飞机出租给航空公司有偿使用的形式,具有融资、融物双重职能。

租赁期满所有权归承租人所有,租赁期限一般10～15年。

租赁公司承担100%的长期借贷风险,租赁公司的租金收入就是息差和利差,还要考虑计提坏帐准备。

折旧方法也是采用直线法,折旧年限是根据租赁期限来确定一般为10年,残值率按30%～40%计提。

融资租赁飞机,租赁公司核定租金涉及的增值税、关税、营业税,如果是GECAS等境外租赁公司还将有预提所得税,不同之处在于租赁公司可以按每期收到租金来分期交增值税和关税,从而将优惠转嫁到航空公司。

6.3　经营租赁

经营租赁提供飞机短期使用权,租期较短,一般5～7年,短的仅半年。租金较高,有较大灵活性。承租人为满足经营上需要融物。

租金的核算可以是固定租金或浮动租金,全新飞机每月租金一般相当于其价值的0.75%～1.5%。经营租赁的租赁费率总是高于融资租赁成本。

实际租金 ＝ 飞机的基本租金 ＋（实际利率 － 预设利率）× 利率差的单位调整系数

保证金支付数量和方式,通常相当于 3 个月的租金的履约保证金。

国外租赁公司租赁飞机享有除使用权以外的飞机所有相关的权利,还包括产权、保险权益、抵押权和留置权等,所有风险由国内航空公司和提供担保的银行承担。

7 总结

航空公司飞机引进方式综合评估策略取决于航空公司的运营策略、现金流状况和融资需求等各方面因素。因此要测算出采购和租赁飞机的资金平衡点,当飞机所有权成本大于平衡点时考虑租赁飞机,反之则考虑采购飞机。

飞机引进最优方式的选择是一项复杂的系统工程,仅仅依靠经济评价做决策是不够的,还必须从公司经营战略、公司财务状况、公司面临的风险等诸多方面进行分析比较,方能做出科学的决策。

参 考 文 献

［1］ Paul Clark. Buying the big jets: fleet planning for airlines (Second Edition), Chapter 6 aircraft Economics ［M］. Ashgate Publishing Limited, England, 2007, p171 - 210.

［2］ 乔治·拉德诺蒂. 航空运输赢利策略［M］. 北京:中国民航出版社,2004:7 - 48.

［3］ 彼得·莫雷尔. 航空公司财务管理［M］. 北京:中国民航出版社,2009.

［4］ Wayne Burns J W. Aircraft Cost Estimation Methodology and Value of a Pound Derivation for Preliminary Design Development Applications ［C］. 53rd Conference of Society of Allied Weight Engineers. Long BeachCA, 23 - 25 May 1994.

［5］ 谭向东. 飞机租赁实务［M］. 北京:中信出版社,2012.

民机复合材料结构设计成本驱动研究

郑 华[1] 赵 荃[1]

（1. 上海飞机设计研究院，上海，200232）

摘要：民机的经济性是关乎其能否获得商业成功的重要因素。在民机复合材料结构工程领域，减少飞机研发及使用阶段的综合成本已成为主要的设计驱动之一。本文面向民机工程，以复合材料平尾壁板作为成本分析模型，按照不同的成型工艺及流程，量化评估了制造成本，并为平尾壁板设计的工艺选择提供成本驱动。

关键词：民机；复合材料；成本；设计驱动

The Study of Cost Driver in Composite Structural Designing of Commercial Aircraft

Zheng Hua[1] Zhao Quan[1]

（1. COMAC Shanghai Aircraft Design and Institute，Shanghai，200232）

Abstract：The economics is the very important factor of business success for a commercial aircraft project. In commercial aircraft engineering, cost reduction during the period of development and service is becoming one of the prime design driver. This paper faced the commmercial aircraft engineering, gave an illustration of the methodology for cost approach employed a panel of composite HTP torque box as an analyzed model, evaluated the fabrication cost in quantity based on the different manufacturing processes, consequently, gave the cost driver as a result for manufacturing selection.

Key words：Commercial Aircraft；Composite；Cost；Design Driver

随着碳纤维复合材料(CFRP)在民机结构中的广泛应用，出现了以波音 B787 为代表的先进民用干线飞机进入航线服役。据统计，B787 飞机全机结构中复合材料用量占 52%，而空客公司正在研发的 A350 飞机也将达到该水平。

复合材料的广泛应用使飞机结构更轻、更坚固。对于民机而言，减重使飞机的使用成本显著降低，这将为航空公司带来巨大的商业利益。但是与金属结构不同的是，复合材料结构的设计与工艺紧密关联。对于同一个结构而言，不同的设计方案将导出不同的工艺流程，而不同的工艺流程所产生的制造成本也不相同。因此，复合材料结构的制造成本受设计和工艺方案的影响更为显著，而合理控制复合材料结构的成本，以成本作为主要设计驱动之一，是保证民机经济性要求的关键。

民机复合材料结构的制造成本主要包括：物料成本、模具成本、人工成本、设备成本等。影响成本使其产生变动的因素称之为"成本驱动因素"，包含：批生产能力、工艺方法及模具投入、结构尺寸及复杂度、人工及设备能力、报废率等。本文通过综合评估民机复合材料平尾壁板基于不同工艺流程的成本及成本驱动因素，为结构设计选型提供决策依据。

1 复合材料平尾壁板模型

民机水平安定面翼盒布置采用多肋式布局，左右翼盒沿航向对称面对接安装。翼盒沿展向布置有前、后梁，翼梁之间沿弦向布置肋，上、下蒙皮为复合材料整体加筋壁板构型，如图1。

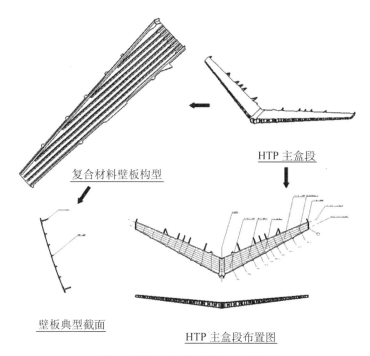

图 1 民机复合材料平尾壁板构型

1.1 结构设计流程

飞机结构设计包含：概念设计、初步设计、详细设计三个阶段。主要设计驱动包含：载荷、结构力学性能、重量、功能性要求、工艺方案、成本等。在概念设计及初步设计阶段，首先应根据外载荷的类型及分布，明确结构传力路线，粗估结构中作用于各元件中的内力，完成结构的布置和传力路线的建立。

其次，针对复合材料结构设计的专业特点，合理选择各结构元件的截面构型、几何尺寸及工艺方法，以结构重量最轻为目标，开展设计权衡分析。在此阶段，需对不同工艺方法进行成本综合评估，引入成本驱动因素分析，评估结构减重优化的成本效益。

　　具体针对复合材料平尾壁板而言,首先对壁板进行分析,完成壁板的结构布置。在飞机巡航状态,平尾主要产生向下的负升力以配平飞机的俯仰力矩,实现飞机的俯仰稳定性和操纵性功能。向下的负升力使得平尾产生向下的弯曲变形,因此,平尾的上、下壁板分别作用轴向拉、压载荷。因此,在壁板上沿展向布置有一定数量的长桁,可以提高蒙皮的稳定性,增加平尾翼面的总体抗弯刚度,提高平尾翼面的结构效率。

　　在此基础上,选择长桁的截面构型,包括:"T"字型、"工"字型、"J"字型、"帽"型长桁等,如图 2。

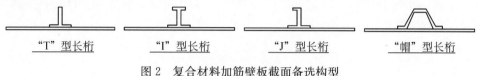

"T"型长桁　　　　　"I"型长桁　　　　　"J"型长桁　　　　　"帽"型长桁

图 2　复合材料加筋壁板截面备选构型

　　考虑单通道民机平尾的载荷水平远低于机翼、机身等主承力结构。因此,为避免讨论的复杂性及结论的不确定性,本文选择常规的"T"字型截面的长桁作为研究对象(以下简称:"'T'型长桁")。

　　采用"T"型长桁的壁板可选择:共固化(co-cured)、共胶接(co-bonded)、二次胶接(secondary bonding)三种常规的制造工艺方案,如图 3。

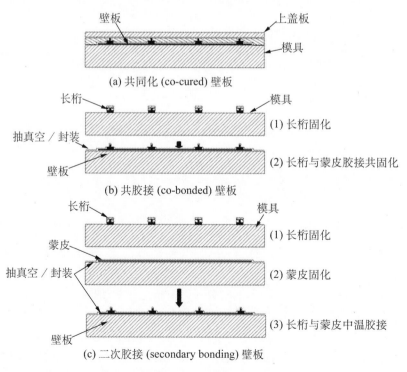

壁板　　　　　　　　　　　上盖板

模具

(a) 共同化 (co-cured) 壁板

长桁　　　　　　　　　模具

(1) 长桁固化

抽真空／封装

(2) 长桁与蒙皮胶接共固化

壁板

(b) 共胶接 (co-bonded) 壁板

长桁　　　　　　　　　模具

(1) 长桁固化

蒙皮

(2) 蒙皮固化

抽真空／封装

(3) 长桁与蒙皮中温胶接

壁板

(c) 二次胶接 (secondary bonding) 壁板

图 3　复合材料加筋壁板模具及固化流程

1.2　制造工艺流程

1.2.1　共固化工艺

共固化(co-cured)工艺是将长桁、蒙皮通过复杂的组合模具铺贴、定位，并一次进罐固化的壁板成型工艺方法。

其中，长桁加热预成型(hot drap forming)工艺的实施方案是：首先在模具表面按长桁铺层铺贴平面，然后通过加热并采用模具使平面的铺层按设计的截面预成型，且使之具备一定刚度，以便在组合模具中定位、封装。共固化工艺流程如图4。

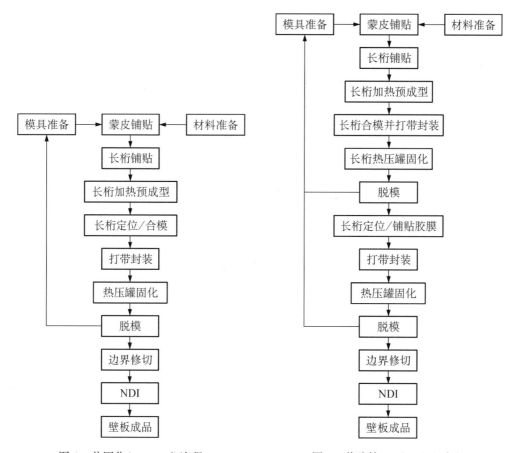

图 4　共固化(co-cured)流程　　　　图 5　共胶接(co-bonded)流程

1.2.2　共胶接工艺

共胶接(co-bonded)工艺是将长桁首先固化后，再与蒙皮通过胶膜胶接共固化的工艺方法。其主要工艺流程如图5。

1.2.3　二次胶接工艺

二次胶接(secondary bonding)工艺是将长桁、蒙皮分别固化后，通过胶膜胶接的工艺方法。为防止弱粘接及胶接界面分层等缺陷,依据破损安全设计准则,二次胶接工艺通常需要在胶接界面区域安装紧固件。其具体流程如图 6。

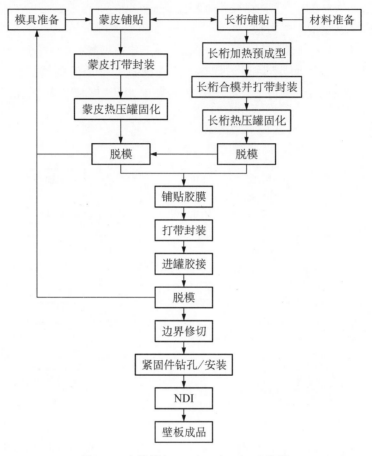

图 6　二次胶接(secondary bonding)流程

2　成本分析

2.1　基于制造工艺流程的成本估算模型

目前,国外产业界提出基于复合材料零件制造工艺流程(Manufacturing Processing-based Cost Modeling,MPCM)的成本估算模型。MPCM 模型面向工程,合理地反映了复合材料结构的尺寸及复杂性,及工艺流程复杂度等成本驱动因素。因此,本文基于 MPCM 模型对复合材料平尾壁板的单架次稳态生产状态下的三种备选工艺的制造成本进行估算。主要步骤包括:①确定成本估算对象;②列出备选工艺流程;③建立每步工艺流程的成本关系;④汇总每步工艺的成本。

为避免讨论的复杂性,本文仅选取单架次平尾壁板的稳态生产作为分析对象,因此,对于文中三种备选工艺中的许多工艺步骤所需的时间可定义为常量,并且仅需对三种备选工艺中的差异流程进行建模分析。

2.2 双曲线模型

双曲线模型适用于广延型工艺及操作,具体表达如下:

$$T = S_t + \left[D_t + n\sqrt{\left(\frac{x}{v}\right)^2 + \left(\frac{2\tau \cdot x}{v}\right)} \right]$$

其中:S_t 定义设备准备时间,包括:模具清洗、探伤,自动铺带机调试等非重复性操作;D_t 定义延迟工时;x 定义结构特征尺寸;v 定义生产速率;n 定义零件数,$n = 1$;τ 定义动态时间因子。

2.3 设备及人工成本

2.3.1 自动铺带工艺

自动铺带的单个工步包括以下动作:机头抬升、切带、压紧、铺带、机头抬升并移动至下一工步起始位置。根据双曲线模型,可将铺带动作视作匀速的过程,其余动作为延迟工时。因此,自动铺带所需操作工时可以表达为:

$$T_\theta = S_t + N_{\text{plies}} \cdot \left[\frac{A}{w \cdot v} + t_{\text{delay}} \cdot \left(\frac{a \cdot L}{w} + \frac{b \cdot A}{w \cdot L} \right) \right]$$

其中:A 定义铺带面积;L 定义零件长度;v 定义铺带速率;w 定义预浸料带宽;t_{delay} 定义每次铺放工步的延迟时间;a,b 分别为铺贴角度常数,定义铺贴角度为 $0°$ 时,$a = 0$,$b = 1$;铺贴角度为 $45°$ 时,$a = 0.707$,$b = 0.707$;铺贴角度为 $90°$ 时,$a = 1$,$b = 0$。

2.3.2 热压罐固化工艺

考虑单架稳态生产,可将热压罐单次使用(预浸料固化及胶膜胶接固化)费用视作常数。

2.3.3 热隔膜成型工艺

热隔膜成型(hot drap forming)工艺无需加压,因此仅需使用烘箱加热。考虑单架次稳态生产,可将烘箱单次使用费用视作常数。热隔膜成型所需模具成本另算。

2.4 模具成本

复材零件的成型模具通常采用 INVAR 钢生产。模具生产的主要工艺流程包括:下料、钣弯、模面机加、焊接、抛光、镀层、检验等。可基于工步法建立成本模型,也可根据经验系数估算。本文选取经验法,依据以下公式进行估算:

$$C_{\text{tool}} = f \cdot x$$

其中,f 为经验系数;x 为特征尺寸。

2.5 制造成本汇总

由于涉及商业机密,本文难以获得完整确认的工艺数据,因此,本文依据所查找的文献数据,对各步工艺的设备准备时间、工步延迟时间及材料使用率等关键参数进行假设。设备及人工成本见表1。

表1　设备及人工成本

工艺步骤	T/h	D_t/h	S_t/h	$1/v/(h)$	工艺参数	
材料准备	1.0	0.5	0.5		材料利用率 η:	0.5
模具准备	3.0	1.0	2.0		零件面积/m^2:	4.874
自动铺带	9.87	1.5	2.0	6.37	预浸料宽度/mm:	150
热隔膜成型	3.33	1.0	2.0	0.33	长桁预成型长度/m:	6.7
打带封装	1.0			1.0	零件周长/m:	16.528
热压罐	7.0		1.0	6.0	—	
合计	25.2					

因此,三种备选工艺方案对应的制造成本分类汇总见图7。

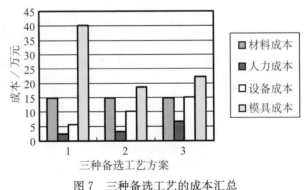

图7　三种备选工艺的成本汇总

1. 共固化;2. 共胶接;3. 二次胶接

三种备选工艺的制造总成本见表2。

表2　复合材料平尾单件稳态生产总成本对比

备选工艺	相对成本对比/%
共固化工艺	100
共胶接工艺	70.3
二次胶接工艺	92.7

2.6　结构设计构型决策

综合以上结果可知:在单架次稳态生产的情况下,共固化壁板制造成本最高,其次是二次胶接工艺,成本最低的为共胶接工艺。主要原因在于共固化工艺所需要的模具为高精度对合模,模具的一次性投入巨大。但共固化工艺的可重复性费用(每架次设备及人工成本)最低,因此,一旦工艺稳定,共固化工艺随批产架次的增加,每架份的成本递减。而二次胶接工艺的可重复性费用最高,因此,批产的成本优势最低。

此外,在综合权衡每架次制造成本的同时,应结合工艺成熟度和稳定性,加入产品报废率的成本驱动因子。共固化工艺属于先进复合材料制造工艺,随着结构尺寸的增加,其缺陷率也成倍增加。即使工艺稳定,依据国内外目前的制造能力,但返工难度大,报废率高。因此,选择共固化工艺将面临一定的成本控制风险。

基于以上结果分析,按照制造成本驱动,复合材料平尾壁板可选用共胶接工艺方案,并据此进一步完成壁板结构细节设计。

3 结论

本文基于民机复合材料平尾壁板的工艺流程,比较了三种备选工艺方案的制造成本,研究了成本驱动对结构设计方案决策的影响。由于难以全面获得准确的工艺参数,本文依据工程经验,在假定参数的基础上,构建了复合材料平尾壁板制造成本的模型,并获得单架次稳态生产的成本数据,为结构设计方案的决策提供依据。由此可见:对于民机复合材料结构而言,成本驱动下的结构设计可以降低项目的成本及风险、强化项目成本控制,使民机项目符合经济性要求。

进一步的工作可依据工艺经验及预研结果,引入定量的产品报废率因子,建立产品批产成本模型,为结构设计及方案决策提供更为基础的依据。

参 考 文 献

[1] G Clayton, C Howe. Cost as a Driver in Advance Composite Aerospace Research [J]. Proc. Instn Mech. Engrs, Vol. 218 Part L: J. Materials: Design and Applications, 2004.

[2] Gutowski T G, Neoh E and Polgar K C. Adaptive Framework for Estimating Fabrication Time of Advanced Composites Manufacturing Processes, October 1995 (Massachusetts Institute of Technology).

[3] R Curran, M Mullen, N Brolly, M Gilmour, P Hawthorne, S Cowan. Cost Modelling of Composite Aerospace Parts and Assemblies [J]. Aircraft Engineering and Aerospace Technology: An International Journal, 2004.

[4] Joshua W Pas (1995). Web Based Cost Estimation Models for the Manufacturing of Advanced Composites, Massachusetts Institute of Technology, Theies of Master Degree.

[5] 叶强,陈普会,柴亚男. 复合材料结构制造成本估算模型及软件开发[J]. 复合材料学报,2008年8月.

民用飞机设计过程中经济性管控研究

曾　征[1]　邬　斌[2]　芦　健[1]

(1. 北京东方慧实科技有限公司，北京，100071)

(2. 中国商飞上海飞机设计研究院，上海，201210)

摘要：本文针对设计对民用飞机经济性起决定性作用，提出在民用飞机的设计阶段通过选择经济性好的方案、制定合理的成本目标、运用挣值管理方法管控研制成本，动态持续地进行交付物经济性分析与控制的方法管控制造和采购成本，运行成本等方法来管控民用飞机的经济性，并利用参数法估算软件 TruePlanning 来支撑这些方法的落地。

关键词：民用飞机经济性；寿命周期成本；挣值管理；TruePlanning

Study on Civil Aircraft's Economics Control during Design Process

Zeng Zheng[1]　　Wu Bin[2]　　Lu Jian[1]

(1. Beijing Eastwise Science & Technology Co., Ltd., Beijing, 100071)

(2. Shanghai Aircraft Design and Research Institute of COMAC, Shanghai, 201210)

Abstract：Design is crucial to military aircraft's economics. In view of this fact, this paper presents some methods to control economics of military aircraft, such as choosing a concept of better economics, establishing reasonable cost targets, controlling development cost by EVM (Earned-Value Management), and controlling manufacture, procurement and operation costs by dynamic and continuous analysis and control of deliverable's economics. In addition, the paper also explains how to apply TruePlanning—a kind of parametric estimating software—to practice these methods.

Key words：Military Aircraft's Economics；Life-Cycle Cost；EVM；TruePlanning

1　民用飞机经济性的重要意义

民用飞机是一种知识密集、技术密集、资金密集的高新技术产品，具有研制成本高、投资回收期长、运营成本占全寿命成本比例大，客户要求苛刻等特点。据统计，研制一种干线飞机机型，要花费 20～50 亿美元，耗用 8～10 年时间，且销售量达到 300 架以上才能达到盈亏平衡。全世界已研制成功的喷气式干线民用飞机中，75%

尚未收回研制成本。另据分析，"四性"（安全性、舒适性、经济性和环保性）是衡量民用飞机市场适应性和发展前景的核心指标，其中经济性是唯一薄弱环节。可见随着民用飞机市场不断变化、全球竞争逐渐激烈、先进技术持续更新，新型民用飞机若要取得商业成功，必须重视经济性问题。

中国商用飞机有限责任公司（以下简称中国商飞）是我国大型民用飞机研制的主体承担单位。中国商飞的首要远期目标就是要取得大型客机和支线飞机项目商业成功。C919 飞机研制是国家经济发展战略实施的一项重要举措，取得商业成功是国人的殷切希望。目前中国商飞 C919 大型客机提出了比同类飞机的油耗少12%～15%，单机价格低 10%，运行成本低 10% 的经济性目标。C919 的竞争目标为波音 737 系列飞机和空客 320 系列飞机。空客 320 目前是全球最畅销的单通道飞机，全球订单已超过 6400 架，现役飞机将近 4000 架，全球客户和用户数量超过300 家。波音 737 享有"世界航空史上最成功的民航客机"的美誉，它也是民航业最大的飞机家族，波音 737 系列的所有机型已获得 6000 多份订单。因此为了在市场竞争中取胜，C919 的经济性是型号成败的关键因素，但要达到 C919 经济性的目标面临巨大的挑战。

2　民用飞机的经济性主要在设计阶段决定

经济性是指组织经营活动过程中获得一定数量和质量的产品和服务及其他成果时所耗费的资源最少。经济性主要关注的是资源投入和使用过程中成本节约的水平和程度及资源使用的合理性。民用飞机的经济性狭义上指针对相同的签派可靠度，即降低飞机的寿命周期成本（Life Cycle Cost，LCC）；广义上还可考虑技术外溢效应、经济拉动效应、产业集群效应等。本文主要研究在设计过程中管控飞机LCC 的方法。根据民用飞机的特点，LCC 由以下几部分组成：

1）研制成本

研制成本包括概念设计、初步设计和详细设计；机体结构和系统的试验和验证；适航验证和试飞；新工艺开发；工装设计和制造等阶段发生的成本。

2）制造和采购成本

制造和采购成本包括原材料、发动机和设备成品的采购；机体制造；飞机总装；产品质量控制和批生产试飞等阶段发生的成本。

3）运行成本

运行成本（Total Operating Cost，TOC）包括运行期间的所有权成本（折旧、利息、保险）、现金成本（燃油、空勤、维修、起降、和导航费）等直接运行成本（Direct Operating Costs，DOC）和机票预订、销售、广告、宣传、行政管理、地面资产、设备的租赁、维修、折旧等间接成本（Indirect Operating Costs，IOC）。

4）处置成本

处置成本指飞机进入处置阶段发生的成本。飞机可能转售或租赁给其他用户运营,或改装成货机,或被封存解体。

从大量实际项目统计得到的帕累托曲线告诉我们:在项目早期的设计阶段,实际成本消耗虽然只占总成本的 30% 左右,但已经决定了 LCC 的 95%,因为在设计阶段就决定了民用飞机的性能指标、规格、总体布局与具体结构,这就从总体上决定了民用飞机的技术参数、生产效率、能耗的大小,可靠性、维修性的优劣以及维修成本的高低,也就基本上决定了民用飞机经济性的优劣。

民用飞机各阶段对 LCC 的影响程度各不相同,而且是不断递减的。一个民用飞机越到后期,改变它的性能、结构、使用条件等固有因素的可能性就越小,等到运营和支持阶段,如要想再减小 LCC 或再去想办法降低运行成本,就基本上不会有太大作为了,见图 1 的曲线。

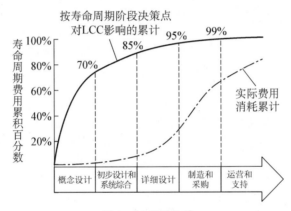

图 1　帕累托曲线

由此可见,设计对民用飞机经济性的影响非常大,可以说对民用飞机经济性起决定性的作用,要管控民用飞机的经济性,主要应该在民用飞机的设计阶段采取相应的措施,并且越早采取措施效果越好。

由于处置成本占寿命周期成本的比例比较小,不是关注的重点,在整个设计过程中,主要通过控制研制成本、制造和采购成本、运行成本,来实现民用飞机经济性的目标。

3　设计过程中民用飞机经济性管控的基本方法——目标成本管理

管控民用飞机经济性主要目的是在满足签派可靠度的前提下降低民用飞机的LCC,并将 LCC 指标作为独立变量贯穿到设计过程中去。在设计过程中管控民用飞机经济性的基本思路是在满足安全性、舒适性、环保性等指标前提下选择经济性比较好的方案;针对具体的方案进行性能和成本的权衡、研制成本、制造和采购成本、运行成本的权衡,确定合理的成本目标;在设计过程中运用挣值管理方法管控研制成本,并动态持续地进行交付物经济性分析与控制的方法管控制造和采购成本、运行成本。做好了这几方面的工作,研制出的民用飞机就既能满足使用的需求,又能将成本控制在目标范围内，最终取得商业成功。

3.1 选择经济性好的方案

在民用飞机的设计过程中，不论是在方案论证阶段、初步设计阶段，还是在详细设计阶段，不论是整体还是各系统、分系统的设计，一般会有多个方案被提出来，多方案选择会贯穿整个飞机的设计过程。为了管控民用飞机的经济性，方案选择非常关键。在多方案选择过程中，需要正对民用飞机的"四性"（经济性、安全性、舒适性和环保性）进行平衡，因为经济性与安全性、舒适性和环保性是相互矛盾相互制约的。正确的做法应该首先筛选出满足安全性、舒适性、环保性、政策法规、机场适应性等基本要求的方案，估算每个方案的 LCC，分析每个方案的效费比，综合分析比较签派可靠度、寿命周期成本以及效费比，选择既能满足客户需求，经济性又比较好的方案。

3.2 确定合理的成本目标

要在设计过程中管控民用飞机的经济性，一方面在设计的前期不仅要制定性能方面的目标，还需要制定成本目标，并将成本作为独立变量贯穿到设计过程中去。在制定性能和成本目标时要做好两个权衡：性能与成本进行权衡，制定出满足使用需求的经济合理的性能和成本目标；从 LCC 角度进行研制成本、制造和采购成本、运行成本的权衡，分别制定研制成本目标、制造和采购成本目标、运行成本目标，保证寿命周期成本总和最低。

1）成本与性能的权衡

民用飞机的性能和成本是相互联系的，性能-成本关系曲线（图 2）告诉我们：随着性能要求的提高，成本会随之增加，但增加到一定值，性能略微提高，成本会大幅度提高。因此我们在民用飞机研制早期，尤其是在方案阶段进行方案的论证与验证、制定性能目标过程中，一定要有成本意识，不要不顾成本片面追求高性能，也不能片面地压低成本，要在性能和成本之间平衡，综合考虑性能、成本、效费比等多种因素的情况下选择技术路线和技术实现途径，制定民用飞机的性能目标和成本目标，以取得最大的投资回报。

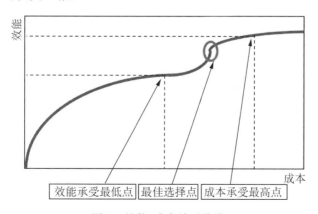

图 2 性能-成本关系曲线

2）研制成本、制造和采购成本、运行成本的权衡

民用飞机的研制、生产、使用是民用飞机全寿命周期的不同阶段，一般也是由不同单位负责，这样在研制和生产过程中各个不同单位为了追求部门利益就可能尽力去降低研制成本、制造和采购成本。可是民用飞机研制成本、制造和采购成本、运行成本之间具有一定的内在关系，是相互影响的。片面追求低研制成本，可能导致制造和采购成本、运行成本大幅增加；同样片面追求低制造和采购成本，可能导致运行成本大幅增加。

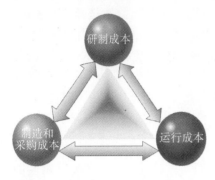

图 3　研制成本、制造和采购成本、运行成本权衡

知道了民用飞机研制成本、制造和采购成本、运行成本这种相互影响的关系，我们在制定民用飞机的成本目标时就要从民用飞机 LCC 的角度系统性地考虑和控制研制成本、制造和采购成本、运行成本，在三者之间平衡，合理地制定这三者目标值，保证 LCC 总和最低，保证民用飞机的经济性。

3.3　设计过程中经济性分两个层面管控

到了初步设计和详细设计阶段，将民用飞机 LCC 控制在目标范围内成了经济性管控的主要任务。在此阶段民用飞机 LCC 分为研制成本和研制出的飞机的经济性两个层面，每个层面由于特点不同应采用不同的方法进行管控。

从设计过程来看，研制成本是飞机设计和研制过程中发生的成本，是实实在在会发生的成本。而制造和采购成本和运行成本是针对设计和研制出来的飞机而言的，是根据设计阶段的交付物对将来会发生的成本的估计值。这两者有本质的区别，管控的方式也截然不同。因此我们将设计过程中的民用飞机寿命周期成本管控分为对应的两个层面：研制成本管理和飞机经济性分析与控制，如图 4 所示。

图 4　设计过程 LCC 管控的层面

3.4 研制成本管控——挣值管理

民用飞机研制具有项目的所有特征,因此研制成本管理完全可以按项目成本管理的思想进行管理。项目成本管理最有效的管理方法是挣值管理(EVM)。挣值管理根据项目的工作分解结构(WBS)以及组织分解结构(OBS)建立项目成本的控制账户。将总研制成本的目标值分解到每个控制账户形成控制账目计划(CAP),结合项目进度计划生成计划价值(PV);根据项目的实际进展计算挣值(EV);根据财务数据和工时情况计算实际成本(AC);基于 PV、EV、AC 从三个维度计算项目的进度偏差(SV)、成本偏差(CV)、进度绩效指数(SPI)、成本绩效指数(CPI),对项目的进度和成本进行准确评价。并进一步分析预测项目的完工时间、"完工尚需估算"(ETC)、"完工估算"(EAC),项目管理者将根据计算分析结果采取相应措施,高层领导将根据计算分析结果为项目预留合理的管理储备(MR)或应急储备(CR),从而达到对民用飞机研制成本进行精确监控,科学决策的目的,如图 5 所示。

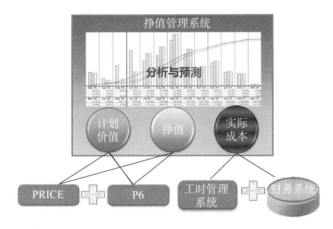

图 5　EVM 示意图

3.5 制造和采购成本、运行成本管控——动态持续进行交付物经济性分析与控制

为了在设计过程中控制交付物的经济性,参考美军定费用设计的思想,在设计过程中首先将整个民用飞机的制造和采购成本、运行成本的成本目标分解细化形成各部段、系统、分系统的成本目标,成本目标与技术要求一并下发给各个部段、系统、分系统设计团队。设计团队在设计过程中针对自己负责的部分同时考虑技术指标和经济性的要求,进行性能与成本的权衡,选择合适的技术方案、技术途径;团队内部针对自己设计成果的成本进行估算,如突破了成本目标,达不到经济性要求及时采取相应措施,直到技术与经济性同时满足要求。型号经济性管控部门在民用飞机研制的各阶段、各里程碑节点根据当前的设计方案反复进行估算,分析估算结果是否在成本目标范围内。必要时在性能、成本、进度间权衡,更改相关设计,或更新成

本目标、技术要求。这样把民用飞机每个组成部分的成本都控制在成本目标范围内,整个民用飞机的制造和采购成本、运行成本就能控制在目标范围之内,保证了研制出的飞机的经济性,如图 6 所示。

图 6 　交付物经济性分析与控制示意图

4 　设计过程中民用飞机经济性管控的利器——TruePlanning

在设计过程中管控民用飞机经济性的各项活动能顺利开展的前提条件是能及时得到准确的研制成本、制造和采购成本、运行成本估计值。准确估算对于在设计过程中管控民用飞机经济性是至关重要的。

但是,要在项目早期的设计阶段实现准确估算不是一件容易的事情,这面临巨大的挑战,因为在设计阶段飞机的技术、经济数据都不太全面,但还要又快又准地完成估算。针对设计阶段的特点及对 LCC 估算的基本要求,类比估算法太过粗略、主观性较强,只适用于概念性和概算性估算;项目早期能提供的信息不足以支持工程法估算;参数法基于技术参数和成本关系式,能快速估算结果,结果较为客观,准确度比较高。因此,在设计阶段采用参数法估算研制成本、制造和采购成本、运行成本,并据此控制飞机的经济性是明智的选择。但由于参数估算法需要基于大量的历史数据来回归成本关系式,但中国商飞是第一次研制大型民用客机,没有历史数据积累。令人欣慰的是美国 PRICE Systems→公司基于 40 余年的大量项目实际数据,已经回归出我们经常会用到的成本关系式,并开发出了参数法估算软件,并已在全球多个国家的政府机构、军工集团及复杂系统供应商中得到广泛应用。我们可以直接将该软件拿来,进行一些针对中国商飞的个性化设置后即可对民用飞机进行全寿命周期的成本和进度进行评估。

4.1 　系统简介

TruePlanning 是美国 PRICE 公司开发的第三代参数法估算软件,可对各种硬件、软件,以及 IT 系统的研制成本、制造和采购成本、运行成本,以及项目进度和风险进行整体估算和分析。

TruePlanning 由 True Systems 系统级估算模块、True H 硬件产品估算模块、True S 软件产品估算模块、True IT 信息化系统估算模块组成。

4.2 系统应用过程

下面以估算动车 LCC 为例，介绍应用 TruePlanning 系统估算 LCC 的主要过程。

1）建立 EBS

在建立估算分解结构（Estimate Breakdown Structrue，EBS）之前，先要确定估算目的和范围，建立 EBS 时考虑因素有估算目的、范围、当前能提供的数据，以及设备的 PBS、研制 WBS、组织分工等。EBS 分解的粗细程度主要根据估算目的、能提供的数据而定，比如初期是否投标的估算可以分解得比较粗，在详细设计阶段为了贯彻按费用设计而建立成本基线，就要分解细一些。EBS 分解的方式主要根据估算范围而定。一般估算研制成本按阶段分解，分解时除考虑 PBS 和 OBS 以外，还要考虑 WBS；估算生产和使用保障费按产品组成分解即可，分解时主要考虑 PBS 和 OBS；估算使用保障费，至少要分解到 LRU。

如果同时要估算研制成本、制造和采购成本、运行成本，要根据设备的具体情况决定共用同一个 EBS，还是分别建立 EBS。

图 7 是民用飞机 EBS 的实例。

图 7　民机 EBS

2）输入估算参数

EBS 建立以后，需要针对 EBS 逐项收集物理、技术参数，并输入到估算模型中，如图 8 所示。

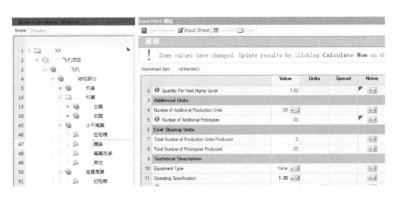

图 8　估算参数录入界面

其中估算硬件的研制成本必须输入的参数有：研制或生产的开始日期、样机数量、运行环境、结构部分重量、电子部分重量、体积、结构部分成本密度、结构部分新设计百分比、结构部分重复设计百分比、电子部分成本密度、电子部分新设计百分比、电子部分重复设计百分比、工程复杂度、结构件外部装配复杂度、电子件外部装配复杂度，所有其他的输入参数可以取默认值。

估算硬件制造和采购成本必须输入的参数有：研制或生产的开始日期、批产数量、运行环境、结构部分重量、电子部分重量、体积、结构部分成本密度、结构部分重复设计百分比、电子部分重复设计百分比、结构件外部装配复杂度、电子件外部装配复杂度，所有其他的输入参数可以取默认值。

估算运行成本需要输入部署、硬件、使用维修三个方面的参数。部署方面需要输入维修方案、四级备件库数量、四级维修车间数量、运行小时数、部署数量等参数；硬件方面需要输入平均故障间隔时间、模块种类数量、零件种类数量、单个设备场所允许离线的设备数量、供应商 LRU 维修成本、供应商模块维修成本、LRU 测试设备成本、基层级 LRU 测试时间、设备级 LRU 平均维修时间、设备级 LRU 平均维修时间、基层级 LRU 平均维修时间、中继级 LRU 平均维修时间、基地级 LRU 平均维修时间、基层级模块平均维修时间、模块生产成本、零件生产成本、运输重量、寿命周期学习曲线、将 LRU 的故障定位到模块的测试设备成本、将模块的故障定位到零件的测试设备成本、LRU 生产成本、参考数量、单位重量的运输成本、单个系统操作人员数量、模块结构控制系数等参数；使用维修方面需要输入单位系统操作人员数量、（每月）运行小时数、直接计入项目管理和监控成本、单位运行小时的燃料和电力成本、平均大修间隔时间、单次大修成本、一次性采购系数、使用维护阶段项目管理和监控系数、使用维护阶段直接计入训练成本、年永久性调动百分比、单次永久性调动成本、基地运营人员数、运行设备替换成本、训练设备维护系数、其他任务设备维修系数、每月任务前及任务后的维护时间、操作人员训练成本、操作人员轮转率、直接计入工业战备完好性成本、后勤工作成本等参数。估算使用保障费需要输入的参数看起来很多，但硬件方面的参数系统可以由系统计算得到，还有一些参数可以采用系统的默认值。

针对以上这些参数，TruePlanning 已经将大量过去项目的经验值内置到系统中，可以通过查表、计算器、甚至直接采用默认值的方式得到。还可以用我们自己过去的类似项目的成本数据校准得到。

3）输出估算结果

输入估算需要的物理、技术参数以后，点击计算按钮，系统就能计算出每一级 EBS 的研制成本、制造和采购成本、运行成本，各级 EBS 的人工费、材料费，以及各级 EBS 的活动的进度、活动和资源的相关成本。并能基于这些估算结果进行各种分析，比如成本比例分析、资源可用性/利用率分析、估算结果风险及置信度分析、投资回报率分析等。

4.3　系统应用

TruePlanning 完全满足设计阶段开展多方案选择、确定成本目标、目标成本管理等活动的成本估算要求，除此之外，在生产和使用过程中还可以辅助制定生产策略、选择维修方案、建立维修车间和备件库、评估和控制供应商价格，全面支撑民用飞机经济性管控。

5　小结

随着市场竞争的加剧，民用飞机经济性越来越重要。本文针对设计对民用飞机经济性起决定性作用，提出在民用飞机的设计阶段通过选择经济性好的方案、制定合理的性能和成本目标、运用挣值管理方法管控研制成本，动态持续地进行交付物经济性分析与控制的方法管控制造和采购成本、运行成本等方法来管控民用飞机的经济性。并利用参数法估算软件 TruePlanning 来支撑这些方法的落地。从而实现民用飞机在研制过程中管控经济性，保证研制出的民用飞机在商业上获得成功。

参 考 文 献

［1］　韩景偶.航空民用飞机寿命周期成本与经济分析［M］.北京:国防工业出版社,2008.
［2］　刘晓东.民用飞机寿命周期成本分析与控制［M］.北京:国防工业出版社,2008.
［3］　罗云,张俊迈,吴奕亮.设备寿命周期成本方法及其应用［M］.北京:海洋出版社,1992.
［4］　叶叶沛,李晓勇.商用飞机全寿命成本分析方法［J］.大型客机设计制造与使用经济性研究［M］.上海:上海交通大学出版社,2011.
［5］　叶叶沛,李晓勇.民用飞机经济性设计［J］.航空公司运营经济性分析与飞机设计［M］.上海:上海交通大学出版社,2012.
［6］　美国审计署.估算和管理项目成本的最佳方法［M］,2007.

国产民用大型客机全寿命周期成本评估指标体系研究

陈晓和[1]　翟晓鸣[1,2]

（1. 上海财经大学财经研究所，上海，200434）
（2. 上海财经大学国际教育学院，上海，200433）

摘要：控制全寿命周期成本是提高国产民用大型客机竞争力的重要一环。以往我国飞机研制靠政府投资，利润提取模式是成本的固定百分点，很不经济。如今民用大型客机研制，采取的是市场化、国际化运作的模式，因此，不讲经济效益不行。本文根据波音、空客等公司发展民用飞机的经验，并结合我国实际，建立了国产民用大型客机全寿命周期成本评估指标体系，试图以此为我国民用大型客机全寿命周期成本评估提供可靠依据。

关键词：民用大型客机；全寿命周期；成本评估；指标体系

Research on the LCC Assessment Index System of China's Commercial Big Airplanes

Chen Xiaohe[1]　Zhai Xiaoming[1,2]

（1. Institute of Finance and Economics, Shanghai University of Finance and Economics, 200434, China）
（2. International Education College, Shanghai University of Finance and Economics, 200433, China）

Abstract: Control of life cycle cost is to improve the domestic civil airliner competition ability is important one annulus. Previous aircraft development in China, relying on government investment, profit extraction pattern is the cost of the fixed points, very economic. Nowadays, civil airliner development, take the market, the internationalization of the operation mode, therefore, do not tell economic benefits. According to Boeing, Airbus and other companies to develop civil aircraft experience, combined with the reality of our country, build domestic civil airliner cost of whole life cycle assessment index system, in an attempt to our country civil airliner evaluation of life cycle cost and provide a reliable basis.

Key words: Commercial Airplanes; Big Airplanes; LCC, Assessment Index System

研制和发展民用大型客机，是建设创新型国家的重大战略举措，对于提高我国的自主创新能力和核心竞争力,促进航空工业的跨越式发展,满足快速增长的民用航空市场需求,推动经济和科技的快速发展,皆具有十分重要的意义。

飞机型号研制具有技术含量高、时效性强、周期长、风险大、耗资多、设计层次高和任务艰巨等特点。以往,我国飞机研制全靠政府投资,设计与生产脱节,风险由国家承担,飞机设计研究以满足上级要求为根本,利润提取模式是成本的固定百分点,没有发挥市场机制的作用,从而造成了飞机全寿命周期成本观念不强,也没有专门的人员和机构对飞机的市场需求和运营成本展开有效的分析研究,经济性等因素在设计过程中考虑较少(见图1)。

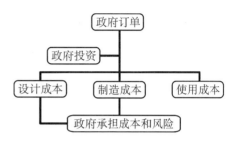

图1　以往我国的军品成本管理模式

如今,国产大型客机的研制和发展模式已不同于以往,它是军民两用技术的有机融合,是传统军工产品研发体制与市场运作机制的有机融合,是自主创新与国际合作的有机融合(见图2)。在经济全球化的背景下,航空工业的全球化是以武器系统的单一研制模式,转向国际化的研发、设计、生产以及市场营销为基础的军民两用研制模式。而且随着经济全球化深度和广度的不断拓展,民用飞机产业的国际竞争越来越激烈。以国产大型客机为代表的我国航空工业要在国际竞争中赢得一席之地,就必须改革现行管理体制和运行机制,确立以成本、收益为核心的经济观念,及时组织开展国产大型客机的全寿命周期成本研究,并建立相应的评估指标体系。

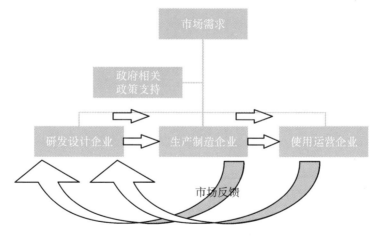

图2　现阶段我国民用飞机研制和发展模式

1　国内外研究现状及评述

根据世界两大民用大型客机制造商——波音和空客公司对下一代大型客机关

键指标的预测,缩减研制周期和研制成本、降低制造成本和控制运输成本都不约而同地成为其战略目标[1]28—33。可见,成本评估指标已经成为决定未来公司命运的关键指标(KPI)。

其中,波音公司在确保飞机安全性的前提下,通过对气动技术、材料技术和系统技术的改进,减轻了飞机重量,降低了耗油率、材料造价和维护维修费用,从而达到了降低飞机全寿命周期成本[1]28—33的目的。同时,通过改良推进技术,降低了推进系统噪声、减少了排放物污染、提高了能源利用效率,从而可以有效规避将来的噪音税、碳排放税、环境税等多项税收,控制飞机的成本(见图3)。

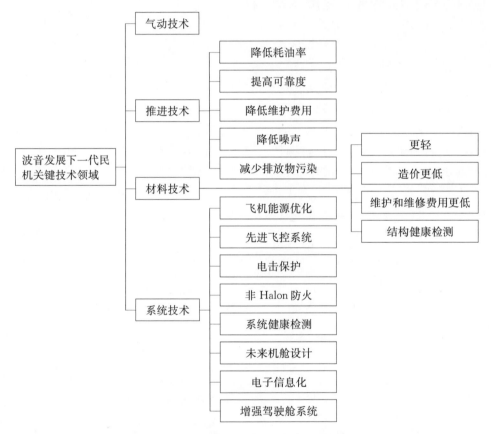

图3　波音公司发展下一代民用飞机的关键技术

空客公司发展民用大型客机的五大战略目标也明确地反映了更安全、更经济、更舒适和更环保的设计理念。它将控制和降低全寿命周期成本放在突出位置[1]28—33,通过更经济的设计、更低的材料和制造成本、更高的燃油效率和飞机使用效率,以及更低的噪声和排放来降低其全寿命周期成本,试图提高其研制生产的飞机在国际市场上的核心竞争力(见图4)。

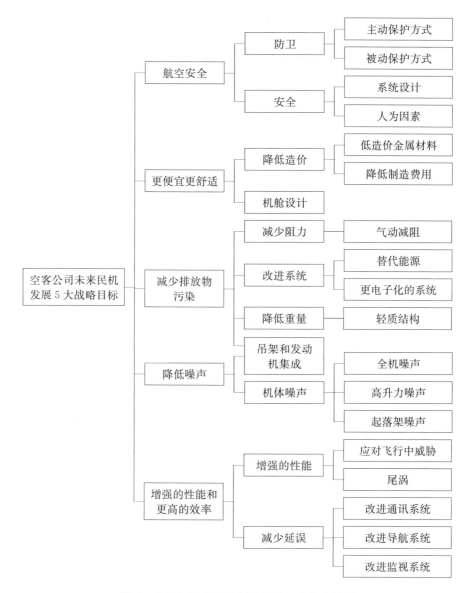

图4 空客公司发展民用飞机的5大战略目标

此外,近几年来在欧美国家政府和行业协会研究发展民用飞机的计划中,降低全寿命周期成本的关键技术研究也成为不变的主题[1]28—33。例如,"AWIATOR"、"EUROLIFT"、"OPERA"等研究计划主要是通过提高大型客机的气动性能来减少飞行阻力,提高巡航效率,降低燃油成本;"TANGO"、"CAI"、"ALCAS"等研究计划主要是通过研究新型材料和合理结构来减轻飞机重量,降低制造成本和燃油成本;"EEFAE"、"UEET"、"ZEA"、"AEAP"、"QGT"等研究计划主要是通过降低飞机排放来提高环保性,降低未来政府对航空运输行业的碳排放、噪音和环境污染等问题的立法所带来的税收成本;"SFW"、"IFCS"、"ACEE"等研究计划主要是通过优化总体设计、更先进的电传技术和人机环境技术等来提高飞机的操作性能,降低维修、地面支援和飞行员的培训成本(见表1)。

表 1 近几年欧美国家重大民用飞机研究计划

序号	研究计划	关键技术	成 效
1	AWIATOR、EUROLIFT、OPERA	提高飞机气动性能,减少阻力,提高巡航效率	降低燃油成本
2	TANGO、CAI、ALCAS	新型材料技术,并使结构更趋合理	减轻飞机重量,降低燃油成本
3	EEFAE、UEET、ZEA、AEAP、QGT	改善飞机的环保性能,降低排放指标	降低排放所带来的税收成本
4	SFW、IFCS、ACEE	总体优化设计、电传技术和人机环境技术	降低地面支援和飞行员培训成本

由此可见,成本指标是提高民用大型客机核心竞争力的关键指标。国产大型客机的研制尚处于起步阶段,以中国商用飞机有限责任公司(以下简称中国商飞,COMAC)所研制的 C919 为代表的国产大型客机正朝着全球民用大型客机 ABC 三分天下的目标迈进。目前,国内这方面的研究多以总结和归纳国外飞机成本评估方法为主,研究适合中国特色以及国产大型客机发展环境的成本评估指标体系的文献尚不多见。因此,本文想在这方面做一些尝试。

2 国产大型客机成本评估指标体系的建立

飞机全寿命周期成本(LCC)的概念起源于军品,其解决方案之一的并行工程设计思想也被首先用于军机[2]31—37 [3]117—134。美国兰德公司(RAND)的 DAPCA(Ⅰ-Ⅳ)[4]354—389模型为我们打开了面向成本的设计思路。但军机研制不同于民机,就是民机企业,经济环境、管理体制和文化背景不同,企业的状况及成本也会有所不同[5]45—48。但是,我们可以按照 DAPCA 模型和波音、空客等公司对民用大型客机进行成本评估的思路,并结合我国实际,来建立我国国产大型客机全寿命周期成本评估指标体系(见图 5)。

图 5 国产大型客机成本评估指标体系

2.1 研发设计成本

研究表明,研发设计阶段对全寿命周期成本的影响最大[6]188—189。帕累托曲线在经济管理[7]84—87和民用飞机全寿命周期成本评估方面的应用(见图6)证明,项目论证阶段70%的LCC已经被决定;而在方案研究结束时,LCC的85%已经被决定;到全面研制工作结束时,LCC的95%已被决定。而使用、维护阶段的活动对LCC的影响只占1%。因此,帕累托曲线理论表明一个民用飞机型号的全寿命周期成本其80%在整个产品全寿命周期的前20%已被决定。由此可得出,研发设计阶段(包含论证阶段和方案阶段)是控制全寿命周期成本(LCC)的关键阶段,其他阶段的控制作用极为有限。

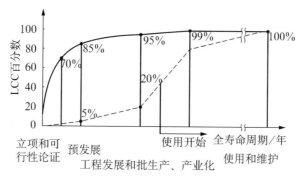

图6 产品全寿命周期各个阶段对LCC的影响

根据帕累托曲线的分析,降低全寿命周期成本应该在民用飞机研发设计阶段下功夫。民用飞机研发设计阶段的成本主要包括研发设计人员经费、仪器设备和设施投入、耗材和物损成本、第三方支持费用等方面(见图7)。其中,第三方支持费用包括为了国产大型客机项目所采购的技术资料、数据和文献等,还包括研发过程中的咨询费以及需要外界提供其他服务的第三方费用。

图7 研发设计成本

2.2 全球采购成本

飞机制造业是典型的全球采购行业,离不开高密度的国际合作。据统计,制造一辆汽车需要 1 万多个零件,而制造一架大飞机则需要 600 万个零件。波音和空客发展至今仍然实行零部件的全球采购,这不但没有削弱它们的地位,反而可以使它们通过整合全球供应商资源使自己变得愈发强大。因此,我国的国产大飞机绝对没有必要每个零部件都自己设计制造,完全可以走"利用全球资源"的道路。事实上,供应商提供的很多零部件不仅质优而且价廉,既有助于我们提高质量,又有助于我们降低成本,还有助于我们获得暂时还不能生产的一些零部件。可见,全球采购其意义十分重大。

目前,民用飞机全球采购的主要模式是"主制造商-供应商"模式。根据国产大飞机的研制现状,我国民用飞机主制造商的全球采购成本主要包括采购国内外供应商提供的复合材料、航空发动机、电子系统、机载设备、电气系统和标准配件,以及先进的生产技术等产品时发生的费用(见图 8)。

图 8　全球采购成本

2.3　制造成本

制造阶段是企业根据民用飞机的总体设计方案,整合国内外供应商资源,通过科学的技术和管理,对民用飞机实施总装制造的过程。总装制造是国产大飞机制造商的核心工作之一,地位非常重要,对于飞机产品的成本控制其意义也十分重大。国产大飞机的制造成本主要包含以下几个方面(见图 9)。

图 9　制造成本

(1) 自产产品成本。主要包括生产机头、机身、机尾等大件时所需要的零部件、配件、材料和耗材等成本。

(2) 制造劳动成本。主要包括企业管理、技术、制造和后勤等人员的劳动成本。

（3）储存和运输成本。储存成本包括存放飞机零部件和产品的仓库和其他储存设施的折旧费和修理费等。运输成本包括运输飞机零部件和产品的费用。

（4）报废成本。自产产品过程中产生的无法通过返工、返修恢复其预期功效，且不能作为让步处理继续使用的不合格产品的费用分摊。

（5）工装成本。大型飞机的装配是一项技术难度大、涉及学科领域多的综合性集成技术，与飞机的最终质量、制造成本和交付周期密切相关。为了国产大飞机的总装制造，在机械设备、装配技术、加工工艺等方面的投入，即为工装成本。

2.4 试飞和适航取证成本

在总装制造完成后飞机即进入试飞和适航取证阶段。试飞是测试和体现飞机性能，获得市场认可，并顺利取得适航证的必要环节。为保证飞行安全、维护公众利益，包括民用飞机在内的各类航空器必须经过上百次的试飞，经受各种不同的飞行状况和操作环境的考验，通过国家适航管理部门的审定并取得适航证后才能投入市场运营。因此，试飞成功并取得适航证是民用飞机进入市场的必要条件。国产大型客机在取得我国的适航证后，还必须取得欧洲和美国的适航证才能够打入欧美市场。取得欧美的适航证是一项复杂而又艰巨的工作。一旦取证受阻，国产大型客机只能在国内销售和运营，这对我国航空制造业和航空市场是一种巨大的伤害。因此，对试飞和适航取证必须给予足够的重视。

试飞和适航取证成本包括试飞成本和适航取证成本两部分（见图10）。其中，试飞成本主要包括试飞期间所产生的飞行员、保障人员、燃油、部件材料损耗、保险费等费用。适航取证成本主要包括在概念设计、要求确定、符合性验证计划、实施符合性验证、证后管理等五个取证阶段所发生的费用。

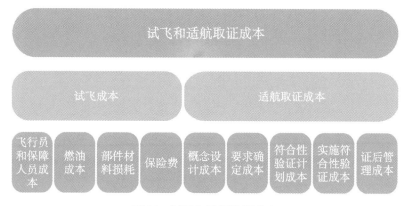

图 10 试飞和适航取证成本

2.5 市场营销成本

即使解决了技术、制造和适航取证难题，国产大型客机依然需要逾越销售障碍。一种成功的机型，除了有良好的安全性和经济性外，还要有规模生产和足够的海内外订单。只有这样,才能把巨额的研发费用分摊在每一架飞机上。国产大型客机的发展是一个循序渐进的过程,需要有一定的市场销售来支撑。依照国际经验,一种新型号飞机,其销售的"盈亏平衡点"一般在300~400架之间。以波音梦想飞机787为例,在签了近480架订单之后其生产的飞机才可以获得利润。因此,对国产大型客机的营销环节必须要有足够的资金、技术和人力投入。

国产大型客机市场营销成本主要包括建立品牌形象、扩大营销网络、形成和维持营销团队、实施促销手段等方面发生的费用(见图11)。其中:

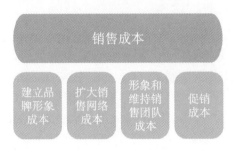

图11　市场营销成本

(1) 建立品牌形象成本。要打破国际民用大型客机现有的AB格局,必须树立起国产大型客机的品牌形象,只有这样,才有可能在销售市场上形成ABC的格局。建立品牌形象,需要有专门的品牌营销策划,如参加各类航空展,邀请全球潜在客户观摩飞行表演,赞助或冠名大型国际活动,组织有影响力的公益活动等。

(2) 扩大营销网络成本。强大的营销网络能够快速将产品投放到目标市场,对于产品销售的重要性不言而喻。国产大型客机既可以借用已有的航空制造企业的销售渠道,也可以通过全球并购获得国外成熟的营销网络,还可以根据市场目标建立营销网点或网络。

(3) 形成和维持营销团队成本。在营销过程中营销人员至关重要,他是市场策略和营销方案的最终执行者,其作用就相当于"临门一脚"。因此,招募、培养和维持一批高素质的营销人员,形成富有战斗力的营销团队,对尽早达到销售的"盈亏平衡点"具有重要作用。

(4) 促销成本。促销是新产品打入市场常用的手段。空客在打入美国市场时用"先飞后买"的促销手段,取得了巨大的成功[8]4-9。国产大型客机在打入国际市场时也会采用一定的促销手段,产生一定的费用,这就是促销成本。

2.6　客户服务成本

一架飞机的使用寿命大约30年,因此民用大型客机不仅需要有高质量的产品,还需要有良好的售后服务作保障。潜在客户选择哪个飞机制造商甚至何种机型直接与飞机售后服务有关,可以说民用大型客机市场已经进入用客户服务来赢得和巩固市场的时代。客户服务成本主要有(见图12)。

图 12　客户服务成本

（1）MRO 网点建立和维持成本。MRO 网点是民用飞机制造商建立的能够提供产品 Maintenance（维护）、Repair（维修）和 Operation（运行）的场所。MRO 网点的建立和维持成本主要包括建设场地、设施和人员，以及提供上述服务所需的成本。

（2）航材和备件服务成本。航材和备件服务关系到民用飞机制造商能否及时、有效的实施产品支援，因此合理的储备与调度航材和备件体系可以增强为客户服务的能力。航材和备件服务成本就是在计划和实施航材和备件服务的过程中所产生的费用。

（3）培训和技术支持成本。民用飞机制造商不仅要销售飞机，还要满足航空公司的合理需求。无论飞机是航空公司从制造商那里直接购买的还是从租赁公司租借的，这种服务都应该贯穿于飞机交付到飞机运营的整个过程。重视客户服务的民用飞机制造商，不仅要供应飞机，还要提供优质的客户培训和支援服务，与各航空公司结成战略伙伴关系。

（4）技术出版物成本。飞机技术出版物是飞机产品客户服务的重要组成部分，是用户决定是否订购飞机的关键因素之一。在飞机交付后的 30 年寿命期内，技术出版物是保证飞机正常运营的重要技术保障。因此，完善、优良的技术出版物，不仅能提高维护、保养和修理质量，缩短排故周期，减少停场时间，保证持续适航和飞行安全，提高用户的经济效益，还能提高飞机的市场形象和制造商的声誉。

2.7　管理成本

管理成本可能是最容易被忽视但又举足轻重的部分。纵观我国民用飞机发展历程，前有"运十"的研制基础，中有为波音、空客两大航空制造业巨头生产多年零部件的技术底子，现有支撑"全球采购"的资金实力，可以克服科学技术和资金方面的障碍。真正缺乏的是管理经验，即对飞机生产线的总装和系统集成，以及对全球海量供应商的管理经验。可以说，管理将成为国产大型客机发展的瓶颈。而且管理经验又非高价引进可以一蹴而就，必须依靠长期实践积累。空客公司就曾因管理不善，导致 A380 推迟交付，引发"空客危机"。目前国产大型客机 C919 的订单形势良好，但如果不重视控制管理成本，就会重蹈空客覆辙。

3　结语

　　长期以来,我国的飞机研制走的都是由政府投资并承担风险的"军品研制-政府采购"模式,全寿命周期成本管理和评估体系处于缺失状态,这显然不能适应国际民用飞机市场的激烈竞争。国产大型客机项目是新时期改革开放和建设创新型国家的标志性工程,肩负着我国工业全面升级的重任,也承载着我国大飞机早日在全世界翱翔的梦想。因此,必须打破原有的军机研制模式,尊重市场经济规律,密切关注并努力降低全寿命周期成本。而要做到这些,就需要牢固树立全寿命周期成本管理的理念,建立相应的成本评估指标体系。

参 考 文 献

[1]　姜澄宇,宋笔锋.从国外民机重大研究计划看我国大型民机发展的关键技术[J].航空制造技术,2008(1).

[2]　徐宗昌.论推行综合寿命周期费用控制法[J].装甲兵工程学院学报,2000,14(1).

[3]　S Castagne, R Curran, A Rothwell, M Price, Æ E Benard, S Raghunathan. A generic tool for cost estimating in aircraft design [J]. Res Eng Design (2008)18.

[4]　G S Levenson, H E Boren, Jr D P Tihansky and F Timson. Cost-Estimating Relationships for Aircraft Airfrmes [R]. A Report prepared for UNITED STATES AIR FORCE PROJECT RAND. 1972.

[5]　党铁红,翟晓鸣.全寿命周期大飞机制造商成本分摊研究[J].国防技术基础,2011(5).

[6]　Asiedu Y, Gu P (1998). Product life cycle cost analysis: state of the art review [J]. Int J Prod Res 36.

[7]　王顺惠.帕累托曲线在经济管理中的应用[J].华东经济管理,1989(3).

[8]　浦一飞.从空客公司看大飞机研制[J].航空工业经济研究,2009(1).

人民币升值对我国民用飞机制造业的影响及其对策

陈晓和[1]   顾嘉琪[2]   韩 阳[3]

（1. 上海财经大学财经研究所，上海，200434）
（2. 中国工商银行上海市分行营业部，上海，200002）
（3. 星展银行中国有限公司，上海，200120）

摘要：为了避免人民币持续升值对我国民用飞机制造业的负面影响，本文分析了人民币升值幅度、收付款比例、通胀水平、飞机国产化率、贷款利率等影响民用飞机制造业并与人民币升值有关的各种因素，对其可能产生的汇率风险进行了量化研究和评估，并据此提出了防范汇率风险的对策建议。

关键词：民用飞机制造业；人民币升值；动态利润模型

RMB appreciation on China's civil aircraft manufacturing industry and its Countermeasures

Chen Xiaohe[1]   Gu Jiaqi[2]   Han yang[3]

（1. Research of Finance and Economics，Shanghai University of
Finance and Economics，200434，China）
（2. ICBC，Shanghai branch of the business department，200002，China）
（3. DBS bank，China Company Limited，200120，China）

Abstract：In order to avoid the continuous appreciation of RMB on China's civil aircraft manufacturing industry negative effect，this paper analyzes the appreciation，collection and payment proportion，the level of inflation，the aircraft is homebred change rate，loan interest rate effect of civil aircraft manufacturing industry and the appreciation of the renminbi in relation to various factors，the possible exchange rate risk quantitative analysis and assessment，and put forward the to avoid or reduce the exchange rate risk countermeasures.

Key words：Aircraft Manufacturer；Appreciation of The RMB；Dynamic Profit Model

1 引言

发展大型民用飞机产业是建设创新型国家、提高我国自主创新能力和核心竞争力的重大战略举措，也是党和国家领导人以及广大人民群众长期以来的良好夙愿。

新中国成立以来,我国高度重视民用航空工业的发展。

2006 年 2 月,国务院将大型飞机项目列入《国家中长期科学和技术发展规划纲要(2006～2020 年)》和《国民经济和社会发展第十一个五年规划纲要》。2007 年 2 月 26 日,国务院常务会议指出:自主研制大型飞机,发展有市场竞争力的航空产业,对于转变经济增长方式、带动科学技术发展、增强国家综合实力和国际竞争力,加快现代化步伐具有重要意义。

民用航空工业具有资金投入多、技术难度大、综合性强等特点。对于发展民用飞机产业的技术和资金,我国航空工业界历来十分重视,而对于人民币升值对民用飞机产业发展的影响则关注不够。因此,有必要加强这方面研究。

首先,汇率的变化对航空运输市场需求会产生重大影响。1986～1989 年,英镑兑换美元升值,英国的一些"追逐阳光族"放弃了欧洲的旅游目的地西班牙,转而奔向了美国的佛罗里达。1992～1995 年,英镑兑换法国法郎严重贬值,而西班牙比塞塔和意大利里拉的贬值幅度却不大,故而前往后两个国家的英国游客急剧增加。1997 年,由于亚洲金融危机导致货币贬值,泰国、马来西亚、印度尼西亚以及其他一些亚洲国家的外国游客大大增加。

其次,汇率的变化对民用飞机制造商的财务状况也会产生重大影响。长期以来,美元一直是国际主要计价货币,新飞机订单大多以美元计价,即便是两大民用飞机制造商之一的空客公司也不例外。这在人民币对美元持续大幅升值的情况下,必将对我国民用飞机制造业产生巨大冲击。因此,如何防范汇率风险,避免其对我国民用飞机制造业的负面影响,已成为亟待解决的课题。

2 我国民用飞机制造业可能面临的汇率风险

2.1 汇率风险概述

汇率风险,又称外汇风险,是指经济主体在持有或运用外汇的经济活动中,因汇率变动而蒙受损失的可能性。在美元是国际主要计价货币的今天,非美国的民用飞机制造商皆不可避免地处于汇率风险之中,这种风险是通过汇率波动对生产能力、飞机定价、生产成本、市场格局和订单合同的影响而进一步影响企业效益和未来一定时期的现金流的。

目前,我国实行的是"以市场供求为基础的、单一的、有管理的浮动汇率制"。汇改以来,随着人民币形成机制弹性的增强,人民币的汇率波动有所加大,其对美元的升值五年已达 22%,年均超过 4%。

2.2 人民币升值对我国民用飞机制造业的影响

人民币升值对我国民用飞机制造业的影响主要表现在两个方面:一是对民用飞机制造商的影响;二是对国内民用飞机生产配套企业的影响。

对我国民用飞机制造商来说,人民币升值将使其处于极为不利的地位。因为空客和波音如今占据着民用飞机市场 90%以上的份额,刚刚起步的中国民用飞机制造业要想在这两个寡头垄断的市场中生存下来就必须取得一定的成本优势,而人民币的持续升值会使这一难度进一步加大;继续以原有的价格接单,经营状况会持续恶化;提高飞机价格,大量订单将会流入空客、波音等竞争对手手中。

虽然部分飞机配套产品依赖进口,人民币升值对于降低国外采购成本能起到一定的作用,但是这方面的节省尚不足以弥补人民币升值所带来的整机业务亏损。

对国内民用飞机配套企业来说,人民币升值的负面影响更大。民用飞机制造商从成本考虑,会将更多采购放到国外配套产品上,这将会恶化国内配套企业的经营环境,削弱其竞争力,不利于提高国内民用飞机配套水平。加之,进口飞机配套设备的性能普遍优于国内产品,必将使民用飞机制造商对国外配套设备的依赖程度进一步加深,从而影响我国民用飞机配套业的发展,导致民用飞机产业链断裂,职工及外包工失业增加,造成社会的不稳定。此外,人民币持续升值还会使我国民用飞机制造商的销售收入、利润和国际竞争力不断下降,民用飞机配套产业的发展环境进一步恶化,从而制约我国民用飞机制造业的发展。

3　动态利润模型、模拟试算及敏感性分析

为了能有效度量民用飞机制造企业的汇率风险程度,需要建立一个多变量的动态利润模型,来研究人民币升值对我国民用飞机制造商可能带来的影响。

3.1　动态利润模型的基本假设

(1) 以美元计价的产品的市场价格不受人民币兑美元汇率的影响;

(2) 企业换汇成本为零;

(3) 国际与国内原材料、配套设备采购成本同比涨跌;

(4) 制造过程可划分为若干个阶段。

3.2　动态利润模型的构建

民用飞机制造成本可以分为国内支出和国外采购两部分,其中国内支出又可细分为国内采购及管理费用(包括职工工资)。所有国外采购都采用美元借款支付,须支付外币利息。影响民用飞机价格和成本的因素有:人民币升值幅度;国外采购占总成本的比例;升值前收款比例;国内采购占国内支出的比例;国内外采购的成本变动幅度;管理费用的变动幅度;外币贷款利率变动。

设利润率为 r,价格为 P 美元,人民币汇率升值前的美元兑人民币为 e,则总成本为 $Pe(1-r)$;在总成本中有 $a(0 \leqslant a < 1)$ 部分是国际采购成本,以美元计价,为 $aP(1-r)$ 美元;$1-a$ 部分为国内采购成本,这部分用人民币支付;设在 $1-a$ 部分中有 $\eta(0 \leqslant \eta \leqslant 1)$ 部分是国内采购成本;美元变动前利率为 I,则动态利润模型为:

$$P' = \sum_{i=1}^{n} Pe\lambda_i(1-\beta_i) \tag{1}$$

式中，$\lambda_i (0 \leqslant \lambda_i \leqslant 1)$ 为民用飞机制造商第 i 期的收款比例，β_i 为第 i 期人民币升值幅度。

$$C^1 = \sum_{i=1}^{n} \theta_i a P e (1-r)(1+\phi_i)[1+I+\delta_i](1-\beta_i) \tag{2}$$

式中，$\theta_i (0 \leqslant \theta_i \leqslant 1)$ 为第 i 期民用飞机制造商实际付款的国外采购成本比例，ϕ_i 为第 i 期国内外采购成本变动幅度，δ_i 为第 i 期利率变动幅度。

$$C^2 = \eta(1-a)Pe(1-r)(1+\sum_{i=1}^{n}\phi_i) \tag{3}$$

$$C^3 = (1-\eta)(1-a)Pe(1-r)(1+\sum_{i=1}^{n}\varphi_i) \tag{4}$$

式中，φ_i 为第 i 期管理费用的变动幅度。

$$C' = C^1 + C^2 + C^3 \tag{5}$$

式中 C' 为人民币升值后以人民币计价的总成本；C^1 为人民币升值后以人民币计价的国际采购成本；C^2 为国内采购成本；C^3 为国内管理费用成本。

$$R' = (P' - C')(1-T) \tag{6}$$

式中，P' 为人民币升值后以人民币计价的总价格。T 为所得税。

$$r' = R'/P' \tag{7}$$

式中，R' 为人民币升值后民用飞机制造商的总利润额，r' 为人民币升值后的利润率。

$$R_0 = R - R' \tag{8}$$

式中，R_0 为人民币升值后所减少的利润额，R 为原利润额。

$$r_0 = r - r' \tag{9}$$

式中，r_0 为人民币升值后所下降的利润率，r 为原利润率。

3.3 模拟试算

设一架飞机需 5 年完成，订购方分 5 年分期付款。B737 - 800 的官方报价为 7250 万～8100 万美元。预计 C919 的售价将会远远低于同类机型，大致上市时的定价会低于 5000 万美元，P 以 5000 万美元计价。此外，设 $\eta = 70\%$，管理费用会随着管理效率的提升而降低，因此用负号表示，民用飞机制造商的利润率 $r = 8\%$，国外采购成本比例 $a = 65\%$，高新技术企业所得税为 15%，一年期美元贷款利率为 5%。美元折合人民币汇率以 2009 年末 6.8282 为起点。其他所有相关变量的设定值如表 1 所示：

表 1　不同期的参数设定

	β_i	λ_i	θ_i	ϕ_i	φ_i	δ_i
1 期	0	0.2	0.15	0.05	−0.03	0.0025
2 期	0.02	0.1	0.15	0.02	−0.03	0.005
3 期	0.05	0.15	0.5	0.06	−0.05	0.01
4 期	0.06	0.15	0.2	0.02	−0.07	0.0125
5 期	0.07	0.4	0	0.02	−0.02	0.0125

注:变量 β_i、δ_i 是累计值

利用 EXCEL 进行编程计算可得:当人民币升值幅度达到 7%,美元贷款利率上调 1.25 个百分点,我国民用飞机制造商的利润额将下降 3080.25 万元,实际利润额为 −758.66 万元(利润额为负,不征收所得税),利润率下降 9.1%,实际利润率仅为 −2.3%。在上述情况下加以变化,比如管理费用因管理效率的提高而下降(第 5 期 $\delta_5 = -0.4$),则利润额上升为 420.4 万元,利润率上涨为 1.3%。

为了探讨人民币升值对民用飞机制造商产生的临界状态,需要将多阶段划分为升值前和升值后两个阶段,由此公式可简化为:

$$R'' = (1-T)Pe[\lambda + (1-\lambda)(1-\beta)] - (1-T)Pe(1+I+\delta) \cdot$$
$$[a\theta(1-r)(1+\phi) + a(1-\theta)(1-r)(1-\beta)(1+\phi)] -$$
$$(1-T)Pe[\eta(1-a)(1-r)(1+\phi) + (1-\eta)(1-a)(1-r)(1+\varphi)] \quad (10)$$

不考虑原材料、设备成本的上涨和管理成本下降以及利率等因素,只考虑升值幅度对利润率的影响。设定基本变量如下:

$$P = 5000, r = 0.08, a = 0.65, \lambda = 0.2, \theta = 0.15, T = 0.15, e = 6.8282$$

表 2　不同升值幅度的影响值

	1%	2%	3%	4%	……	26%	27%	28%
利润额	2236.9	2152.3	2067.6	1983.0	……	120.7	36.0	−57.2
利润率	6.60%	6.41%	6.21%	6%	……	0.45%	0.13%	−0.22%

从表 2 不难看出,人民币升值到 27% 以后,民用飞机制造商的利润空间已经很小,一旦升值超过 28%,民用飞机制造商就会出现亏损。

3.4　敏感性分析

为了衡量原材料上涨幅度、管理成本下降幅度、利率变动幅度等因素对民用飞机制造企业利润的影响,需要进行敏感性分析。由(10)式可得各因素的敏感性公式:

$$\partial R''/\partial \beta = (1-T)Pe[\lambda - 1 + a(1-\theta)(1-r)(1+\phi)] \quad (11)$$

$$\partial R''/\partial a = (1-T)Pe(1-r)[\eta(1+\phi)+(1-\eta)(1+\varphi)$$
$$-\theta(1+\phi)(1+I+\delta)-(1-\theta)(1-\beta)(1+\phi)(1+I+\delta)] \quad (12)$$

$$\partial R''/\partial\lambda = (1-T)Pe\beta \quad\quad\quad\quad\quad\quad\quad\quad\quad\quad\quad\quad (13)$$

$$\partial R''/\partial\theta = (1-T)Pea\beta(1-r)(1+\phi)(1+I+\delta) \quad\quad\quad (14)$$

$$\partial R''/\partial\eta = -(1-T)Pe(1-r)(1-a)(\phi-\varphi) \quad\quad\quad\quad (15)$$

$$\partial R''/\partial\phi = -(1-T)Pe(1-r)(1+I+\delta)[a\theta+a(1-\theta)(1-\beta)]-$$
$$(1-T)Pe(1-r)\eta(1-a) \quad\quad\quad\quad\quad\quad\quad\quad\quad (16)$$

$$\partial R''/\partial\varphi = -(1-T)Pe(1-\eta)(1-a)(1-r) \quad\quad\quad\quad (17)$$

$$\partial R''/\partial\delta = -(1-T)Pe[a\theta(1-r)(1+\phi)]$$
$$+a(1-\theta)(1-r)(1-\beta)(1+\phi) \quad\quad\quad\quad\quad\quad (18)$$

为计算各因素敏感度,特设定参数如表 3 所示。计算结果如表 4 所示。

表 3　参数设定值

β	a	λ	θ	η	ϕ	φ	δ
7%	0.65	0.2	0.15	0.7	0.17	-0.2	1.25%

表 4　各因素的敏感度

$\partial R''/\partial\beta$	$\partial R''/\partial a$	$\partial R''/\partial\lambda$	$\partial R''/\partial\theta$	$\partial R''/\partial\eta$	$\partial R''/\partial\phi$	$\partial R''/\partial\varphi$	$\partial R''/\partial\delta$
-59.57	-29.41	20.31	15.10	-34.57	-238.82	-28.03	-190.96

各因素的敏感度分析:

人民币每升值 1%,民用飞机制造商的利润就会损失 59.57 万元;国外采购比例每提升 1%,民用飞机制造商的利润就会减少 29.41 万元。可见,人民币升值所带来的负面影响是相当大的。

与此相反,升值前的收款比例每提高 1%,民用飞机制造商的利润就会增加 20.31 万元;升值前的付款比例每提高 1%,民用飞机制造商的利润就会增加 15.10 万元。从敏感度公式中不难看出升值幅度越大,两者从中获得的收益就越大。

国内支付的采购成本比例每增加 1%,民用飞机制造商的利润就会减少 34.57 万元,与提高国外采购比例所减少的 29.41 万元相比,民用飞机制造商显然会选择负面影响较轻的,从国外采购原材料和设备,从而使国产化率降低。

国内外原材料上涨会使民用飞机制造商的处境更加窘迫,其对利润的影响更大,原材料、设备的采购成本每上涨 1%,就会使利润减少 238.82 万元。

管理水平提高,生产周期加快,对利润的贡献也不可忽视,其每减少 1%,民用飞机制造商的利润就会增加 28.03 万元。

美元货币的借贷成本也是需要考虑的因素之一,因为每提高 1%,民用飞机制造商的利润就会损失 190.96 万元。

为了借助外部金融市场来达到企业避险的目的,下面探讨汇率风险管理的策略。

4 适合我国民用飞机制造业的避险工具及汇率风险管理策略

根据是否使用金融工具,汇率风险管理的策略大致可分为两类:一类是企业根据自身情况和经营特点,通过内部调整进行避险管理。另一类是利用国际标准化金融交易工具来避免或减少风险。我国民用飞机制造业汇率风险管理策略如图1所示。

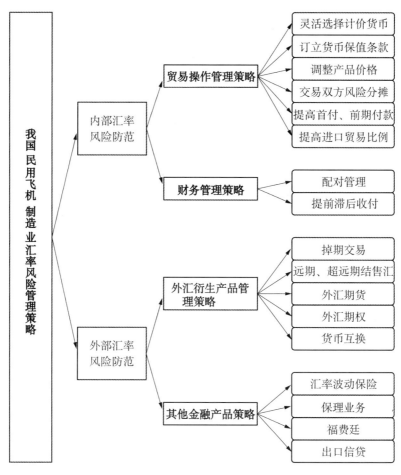

图1 民用飞机制造业汇率风险管理策略

4.1 适合我国民用飞机制造业的避险工具

由于每个策略侧重点不同,其适用范围也不同。为了找到最适合民用飞机制造业的避险策略,需将不同的策略按其工具特性和操作成本进行分析、比较和归纳。

表 5　适合我国民用飞机制造业的避险工具

策　略	工 具 特 性	操 作 要 点
灵活选择计价货币	贯彻"进口支付软货币,出口收入硬货币"和"原币进原币出"两大原则	须得到贸易伙伴的认可,或提供相应的贸易优惠
订立保值条款	贯穿于企业日常业务过程中,效果则取决于企业对汇率走势预测的准确性以及交易方对于交易价格变化的接受程度	须与贸易伙伴进行协商;同时对汇率走势要有相对准确的分析、判断
风险分摊	引入"人民币汇价波动协议",要求双方共同承担风险	合理测算汇率波动对双方的影响
调整首付比例	改善首付款方式	须得到贸易伙伴的认可,或提供相应的贸易优惠
远期结售汇业务	提供一种"收付在未来、汇率当前确定"的服务,从而使得企业在汇率剧烈波动的情况下,及时锁定未来成本和收益	交易时间的选择较为关键
飞机保理	在未来应收账款债权转让和实现融资的同时,改变结汇的时点	应考虑融资成本的影响

由表 5 不难看出:灵活选择计价货币、订立保值条款、风险分摊、调整首付比例、远期结售汇业务和飞机保理是较为适合我国民用飞机制造业的避险工具。

4.2　我国民用飞机制造业汇率风险管理策略

一是加大民用飞机高附加值产品的研发生产力度。由于高端民用飞机产品的利润十分丰厚,汇率的波动虽然对这些企业的盈利会产生一定影响,但是与生产高附加值产品的企业相比影响会相对小一些。

二是通过战略联盟、并购等措施,控制原材料成本,这在人民币升值的背景下尤为重要。这可以通过整合资源,加强原材料采购的谈判能力等来实现。

三是适时推出产业扶持政策。无论是波音还是空客,其发展都离不开政策扶持。政府应该通过减免税负和财政补贴来扶持民用飞机制造业。为了不违背 WTO 的补贴反补贴条款,建议将直接补贴改为间接补贴,将明补改为暗补。

四是选择正确的计价货币。坚持"进口支付软货币,出口收入硬货币"和"原币进原币出"的原则,搞好软硬货币的组合。

五是通过风险分摊防范汇率风险。其过程主要是:确定产品的基价和基本汇率,确定调整基本汇率的方法和时间,确定以基本汇率为基数的汇率变化幅度,确定交易双方分摊汇率变化风险的比率,根据情况协商调整产品的基价。

六是掌握好收付时间。通过出口计价货币坚挺时推迟收款、计价货币疲软时提前结汇等方法来提高企业收益。

参 考 文 献

［1］　崔庆宝. 中国造船企业汇率风险管理研究［D］. 厦门大学，中国优秀硕士学位论文全文数据库，2009.

［2］　李军. 经济模型基础理论与应用［M］. 北京：中国社会科学出版社，2006.

［3］　刘双双，胡庆江. 中国大飞机市场预测分析［J］. 技术经济与管理研究，2008(6).

［4］　苗得雨. 中国民用航空工业金融服务系统研究［M］. 北京：航空工业出版社，2009.

［5］　陶永宏，陈定秋，戈铮. 基于动态利润模型的人民币升值对中国造船业影响的分析［J］. 中国造船，第 50 卷，2009(1).

［6］　汪文飞. 中国航空制造业未来发展的路径选择［D］. 南京航空航天大学，中国优秀硕士学位论文全文数据库，2008.

［7］　Lars Oxelheim and Clas Wihlborg. Measuring macroeconomic exposure：The cas of Volvo Cars［J］. European Financial Management，Vol. 1，1995(3)，P241 - 263.

机载娱乐系统客户化选项功能
及经济性分析

徐 笑[1]

（1. 上海飞机设计研究院市场研究中心，上海，200232）

摘要：飞机产品客户化是飞机制造商帮助航空公司提供差异化服务和提升竞争力的重要环节，机载娱乐系统则是航空公司直接提升客户满意度和丰富旅客乘机体验的重要选项。本文通过对当前主要飞机制造商提供的机载娱乐系统选项的收集和整理，阐述了该系统选项的种类功能、主要供应商和经济性优缺点。通过对比分析其选装时产生的不同经济效益，指导航空公司选装合适的 IFE 系统以减少运营成本和获取增值收益。

关键词：机载娱乐系统；客户化选项；种类功能；经济性分析

In-flight entertainment system customization options' functions and economic analysis

Xu Xiao[1]

（1. The Market Research Center of Shanghai Aircraft Design
Research Institute，Shanghai 200232，China）

Abstract：Aircraft customization enables aircraft manufacturers to support airlines providing different services while enhancing their competitiveness. In-flight entertainment （IFE） system is one of the options to improve passenger's satisfaction and rich their flight entertainment experience. This article expounds different types and functions，main suppliers and economic characteristics of IFE system through collecting and finishing current options. By analyze different options' economic characteristics，guide the airlines to select appropriate IFE system to reduce the operation cost and get the value-added benefits.

Key words：In-flight Entertainment System；Customization Options；Types and Functions；Economic Analysis

1 引言

机载娱乐（In-flight Entertainment，IFE）系统是指航空旅行中在飞机客舱内为旅客提供任何可能的娱乐实现手段的软硬件设备系统，是民用飞机客舱系统的重要组成部分,其性能是旅客判断航空公司服务质量好坏和影响运营成本的要素之一。飞机客户化选项是指用于定义一款特定飞机的性能、功能、设备或服务,它是飞机制造商可提供给客户进行客户化构型配置的项目。目前,IFE 系统已经广泛地被大多数飞机制造商作为一项重要的客户化选项提供给航空公司帮助其实现差异化服务和获取利润。本文针对不同种类的 IFE 系统及其功能、供应商及产品、采购、维修成本和增值空间等问题展开讨论。

2 机载娱乐系统客户化选项概述

客户可用选项是指已设计的、预先定义的、可供客户选择的功能、服务、设备、布置方案、改进的操作和支持等方面的选项,例如发动机、通信导航设备、机载娱乐系统、起落架机轮刹车和轮胎以及座椅、内饰、喷漆等。客户可用选项会影响飞机的性能、功能及价格等,飞机产品客户化对飞机产品的市场开拓和满足不同航空公司营运的差异化需求有着重要的意义。IFE 系统通过与客舱系统互联以及不同媒体服务的结合,实现机上广播、音视频娱乐及购物等多种应用,航空公司通过选择不同的IFE 系统来提高营运的竞争力、服务质量和展现经营特色。正是考虑到不同航空公司客户细分市场的需求以及日益增多的 IFE 种类和不断创新的技术,IFE 系统成为了多数飞机制造商的客户化选项。

IFE 系统目前有两大常用功能,一是通过 IFE 系统选择乘务服务,例如让乘客选择乘务员呼叫、灯光功能等;二是通过 IFE 系统提供娱乐功能,例如旅客自行选择 IFE 中存储的电视、电影、音乐节目功能等。而这些功能的实现方式主要通过壁挂或吊挂式显示屏、椅背式触摸屏及便携式个人娱乐设备等终端设备来实现。航空公司选择不同种类的终端设备会带给乘客不同的飞行体验功能。

3 机载娱乐系统选项的种类及功能

IFE 系统由客舱管理音视频设备、座椅设备以及相互连接的数据总线和供电线路等组成。不同的设备种类提供了不同的功能,形成了面向航空公司的多种客户化选项。

3.1 机载娱乐系统的种类

IFE 系统的客户化选项由可安装在基本型飞机上的可选功能组成,通常 IFE 系统既可以安装在乘客客舱区域,也可以安装在机组休息区域。基本的 IFE 系统会将音频信号和乘客服务功能(登机广播,阅读灯控制)从设备终端传输到每个乘客座位,而个人音视频功能则由安装在每个座椅的旅客控制装置(PCU)控制。机载娱乐系统需要有良好的人机接口界面,并能够简便的进行服务器娱乐内容的更新。目前,主要的 IFE 系统的种类包括了吊装式娱乐系统、椅背式娱乐系统和便携式娱乐系统。

(1)吊装式娱乐系统是在客舱两侧旅客服务装置中间安装吊挂式显示器,间隔数排安装一个。

（2）椅背式娱乐系统是在客舱每个座椅椅背上安装一个触摸屏，乘客通过触摸屏或旅客控制装置实现对各个功能的选择。

（3）便携式娱乐系统是指乘客自带或航空公司提供的手持式娱乐设备。通常情况下，飞机制造商仅提供电源插口功能，在每排乘客的座椅上安装一个或数个可连接电源的插口，供乘客在使用便携式设备过程中维持所需电量。

3.2　机载娱乐系统的功能

目前，机载娱乐系统包括视频、乘客和机组人员通信、乘客信息、互动和乘客个人电源五大功能。

3.2.1　视频功能

视频功能目前分为顶吊式视频系统和个人视频系统两种形式。

顶吊视频系统将视频节目从播放组件传递到显示终端上（DU），例如，顶吊可折叠或者壁挂式 LCD。预先录制的视频节目信号由视频源产生，他们随后通过模块组件，区域分布盒（ADB）和分支装置（TU）发送到顶吊视频系统，而分支装置的数量取决于顶吊显示器数量和他们被分别安装的位置。在飞行期间，乘客可以在吊装显示器上观看视频节目，并且可以借助座椅端的耳机插孔接收该播放视频的音频信号；旅客还可以通过旅客控制装置来选择不同的音频频道。

个人视频系统将视频节目传送到个人乘客座椅终端设备，视频信号是由视频源转换、控制和模块化的。个人视频功能可确保每个乘客选择他们自己喜欢的节目。个人视频通过终端设备向每个乘客座椅提供了视频节目，节目会在每个座椅上安装的座椅显示终端（SDU）里播放。频道选择和声音控制则安装在旅客控制组件（PCU）上。它还可以通过播放器和服务器重新制作预录制节目。

3.2.2　乘客和机组人员通信功能

乘客和机组人员通信功能需要提供给乘客一个电话或者数据传输设备，一个客舱通信装置连结客舱系统和飞机通信转换系统，例如卫星通信（SATCOM）。该功能包含如下系统：

1）乘客个人电话系统

个人电话系统提供给乘客座椅通信设备，将电话功能与 IFE 手持受话机一体化。这是通过 IFE 的网络布置，通过客舱通信装置（CTU）和空中到地面的通信转换系统以及公共地面网络来连结的。

2）壁挂式电话系统

安装壁挂式电话来确保乘客和公共地面电话网络间的通信，这是通过客舱通信组件（CTU），空中到地面的通信系统和公共地面网络来实现的。电话系统可以通过一个专用的网络或者 IFE 网络来连结到 CTU。电话系统的安装位置必须与客户化客舱布局一致。

3）乘客 e-mail 服务系统

这个系统提供给乘客通过 IFE 座位显示设备发送和回复简短的邮件/短信或及

时信息,此系统运用机上 IFE 服务端完成卫星和地面的交流。

3.2.3　乘客信息功能

1)登机音乐和预录制通知

登机音乐和预录制 PRAM 通过客舱交互数据互联系统 CIDS 和 IFE 系统向乘客播放预录制的信息和登机音乐,这个系统由乘务员控制面板(FAP)通过 CIDS 控制,此外,在客舱快速颠簸的情况下会自动播放紧急通知。

2)乘客飞行通知系统(PFIS)

乘客飞行通知系统通过 IFE 视频显示屏不断向乘客提供更新的飞行情况和目的地信息。PFIS 由 IFE 控制面板远程控制,播放下列信息:飞行信息(包括高度,外部温度,地面速度,经度/纬度和飞机位置);飞行路线地图;机场入口地图信息;航空公司 logos;剩余飞行距离;预计到达时间;目的地,抵达和起飞信息等。

3)外部视景照相机

外部视景照相机连接到 IFE 系统上,将提供给乘客高质量的飞机外部视景图像。照相机模块是由平行面视角照相机,垂直面视角照相机和电子设备来生成视频信号并连结到飞机上的。根据起落架状况(放下或收起),外部视景照相机视频频道可以展现平行面视图和垂直面视图,此外,如遇特殊情况,驾驶舱可直接关闭照相机。

3.2.4　互动功能

典型的互动功能一般是玩游戏,收费点播,免税购物和乘客调查等。旅客控制装置/扩展功能的座椅扶手或者触摸屏等都可以让乘客操作这些互动功能,乘客付费时则可以通过使用与座椅一体化的信用卡刷卡机来消费。音频/视频点播功能可以允许个人选择和处理机载娱乐系统服务器上的数字化音频和视频节目。PCU/扩展功能的座椅扶手和触摸屏可以允许 AVOD 节目控制。

3.2.5　乘客个人电源功能

乘客个人电源功能是为旅客手持电子设备提供 110VAC/60 Hz 交流电源,通常情况下,在商务舱内每个座椅上安装一个乘客电源插孔,在经济舱内,考虑到实际使用情况和系统功耗问题,在每排的三个座椅上安装两个乘客电源插孔,该子系统需要有过流保护功能。

表 1　机载娱乐系统的种类及功能

IFE 系统种类	吊装式 IFE	椅背式 IFE	便携式 IFE
视频功能	有	有	有
乘客和机组人员通信功能	无	有	无
乘客信息功能	有	有	无
互动功能	无	有	有
乘客个人电源功能	可选	可选	有

4　机载娱乐系统主要供应商及产品

机载娱乐系统可以根据系统平台的不同,分为两大类:第一类是传统 IFE 平台,基于该平台,音频和视频系统可以独立选择,也可以组合选择,可以选择不同供应商。目前的系统主要是 Rockwell Collins 的可编程音视频娱乐系统(PAVES)和松下公司的多路传输音视频娱乐系统(MPES)。第二类是基于服务器的集成 IFE 平台,该系统是目前 IFE 系统发展的主流,该系统集成了可用于个人视频和手持式电子设备的座椅电源功能。当前的系统主要是松下的 X-Series(eFX)系统和泰雷兹的 Top Series 系统。

4.1　罗克韦尔柯林斯公司

罗克韦尔柯林斯公司的 IFE 设备系列包含了传递预录制信息的吊装式娱乐系统和乘客可自由选择音视频信息的椅背式娱乐系统。该公司开发的主要产品包括 dTES 系统、eTES 系统以及 PAVES 系列。其中,PAVES 系统的核心是高清晰的媒体服务器,其容量大到可存储多个娱乐内容并在变换航段时自动选择内容。新技术开发的 dTES 系统是全数字化的椅背式娱乐系统,可以在单通道客机上创造出环绕立体声的音频效果。

4.2　泰雷兹公司

法国泰雷兹公司的 Top Series 系列娱乐系统是一个可适用于乘客和机组的多功能娱乐平台,支持安装在单通道、双通道和支线客机等多种机型上。该系统提供集成的音视频点播、电源供应和联通功能。其创新的 Top Series AVANT 包含有宽屏显示器、环绕声音效、降噪和高清等功能,可降低播放设备间的相互干扰,具有高可靠性和大容量的特点。该系统还设有座位个人电源插口,乘客可自由连接各种便携式设备。

4.3　松下公司

日本松下公司生产了全系列的机载娱乐系统并提供相应的解决方案,其主打产品包括 eX2、eFX、eXpress 以及数字 MPES 等。X 系列系统可提供网络和通信功能,为乘客提供音视频点播(AVOD)和互动娱乐选项。该系列系统重量轻,使用空间小,可提高飞机燃料效率,减少安装维护设备的运营成本。

5　机载娱乐系统选装的经济性分析

不同种类的 IFE 系统的采购、维护成本和增值项目各有差异,这使得航空公司在选择设备时会根据其目标客户、运营模式和航线等诸多因素选择合适的 IFE 设备,其中,运营成本和增值空间是决定选项的重要因素。

5.1　吊装式娱乐系统经济性分析

运营成本方面,吊装式娱乐系统的终端设备是在客舱两侧旅客服务装置中间安装吊装式显示器,间隔几排安装一个,因此有限的设备数量控制了采购成本。而数量有限的终端显示器降低了安装时设备布线的复杂程度,大大降低了设备的安装成本。与此同时,设备的维修成本也因为设备配置成本较低而有所限制。但是,作为传统的 IFE 设备,设备自身的售价和给飞机造成的额外重量以及燃油消耗的费用都增加了航空公司运营成本。

增值空间方面,吊装式娱乐系统只能简单的播放预录制的音频和视频节目,不能提供给乘客丰富的互动功能,降低了乘客的飞行娱乐体验,也降低了设备增值的空间,在采购成本、安装成本和维修成本都不高的情况下,增值盈利空间也很小。在旅客娱乐需求千差万别的情况下,固化的 IFE 内容的丰富程度和个性化程度显然无法提供给乘客高端的飞行娱乐体验。因此,结合以上经济特征,吊装式娱乐系统通常由低成本航空公司或支线机的经济舱选装。

5.2　椅背式娱乐系统经济性分析

运营成本方面,椅背式娱乐系统需要在客舱每个座椅上安装一个触摸屏,由此产生的与座位数相匹配的设备数量较大,上百个设备的采购成本昂贵。与此同时,这些设备导致的飞机重量增加、燃油费用提高和最大航程的缩短直接影响了航空公司的运营成本。其次,座椅式 IFE 设备安装工作量大,布线繁复,配置成本颇高,采购费用昂贵的同时也要承担着较高的设备安装成本。同时,在乘客高频率的使用椅背显示器的互动功能时,设备的故障率和维修率也相应增加,提高了航空公司的设备维护成本。

增值空间方面,许多航空公司坚信质量好、功能多、可靠性高的 IFE 系统可吸引更多的客源。正是由于椅背式娱乐系统除了具备传统的音视频选择功能,更增加了互动娱乐等功能,使航空公司获得广告收入之外,还可以通过机上通信、购物、点播和游戏等各种应用,不断地挖掘盈利增长点,在提供给乘客满意舒适的飞行娱乐体验的同时获得更多的收益。因此,结合以上经济特征,这种采购维护成本虽高但增值盈利空间巨大的椅背式娱乐系统通常由大型航空公司或干线机选装。

5.3　便携式娱乐系统经济性分析

运营成本方面,由于便携式设备可以由乘客自带,也可以是航空公司提供,无论何种情况下,设备都无须花费时间和费用来通过适航取证。通常情况下,航空公司仅提供由少数配件设备组成的电源插口来帮助乘客维持便携式设备的电量,因此在设备采购成本和安装成本上都降低了许多。其次,便携式娱乐系统所需的个人电源接口大多属于头端设备,较低的配置成本也降低了设备的维护费用。

增值空间方面,便携式娱乐设备即使在仅提供电源的情况下,仍能成为旅客的"私人化飞行娱乐设备",丰富乘客的飞行娱乐体验。但是由于独立于其他机载设备,航空公司无法利用"机载娱乐系统"硬件平台为设备提供乘客信息功能。但是依据 IFE 系统的发展趋势,航空公司在无线客舱技术的支持下,如提供收费的无线网络功能,则能更大程度的发挥便携式娱乐设备的功能性。这样不但丰富了乘客的娱乐功能,也增加了航空公司的额外收入,获得更高的增值收益。因此,便携式娱乐系统的低采购维护成本和潜在的增值盈利点正吸引越来越多的航空公司选装该系统。

表 2   机载娱乐系统选项的经济性优缺点

IFE 系统	吊装式娱乐系统	椅背式娱乐系统	便携式娱乐系统
优点	a) 运营成本(设备采购、安装、维护费用)适中 b) 提供乘客基本娱乐体验功能	a) 互动功能丰富,提供乘客高端的飞行娱乐体验 b) 增值盈利空间大,增值项目多	a) 运营成本(电源设备采购、安装、维护费用)低 b) 无须适航取证 c) 提供乘客私人化娱乐功能 d) 潜在增值空间大
缺点	a) 燃油消耗费用增加 b) 缺少互动功能 c) 增值盈利空间少	a) 运营成本(设备采购、安装、维护费用)高 b) 设备损坏维修率高 c) 燃油消耗费用增加	a) 独立于机载娱乐平台,不能提供飞行信息和通知广播等功能

6   总结

    本文通过对机载娱乐系统客户化选项的种类功能、供应商和经济性特点的分析,对比出三种 IFE 系统选装对航空公司运营成本和增值收益产生的不同影响。结论显示吊装式 IFE 系统适合低成本航空公司和经济舱选装,椅背式 IFE 系统的选装范围目前最为广泛,而便携式 IFE 系统的低成本和高增值的优势将成为今后航空公司选装的趋势。综上所述,航空公司通过选装 IFE 系统为乘客提供丰富多元化的互动娱乐项目,既提升了服务质量也可获得新的利润增长点。与此同时,飞机制造商也可以通过客户选型工作,完善可提供给航空公司 IFE 系统的选项配置,实现互惠共赢。

参 考 文 献

[1] 吴康. 机载娱乐系统发展概述[J]. 通信技术,2012(07):103-105.

[2] 刘跃. 新一代机载娱乐系统进入客舱[J]. 国际航空,2009(08):64-65.

[3] David Li. How to make in-flight entertainment system (IFE) more cost-effective [EB/OL]. (2012-02-09)[2012-12-10]. http://wenku.baidu.com/view/74eab0d3195f312b3169a5fe.htm.

[4] In-flight Entertainment and Cabin Power Configuration Guide. BLMC. 2008(05).

民用飞机价值探讨

张　康[1]　朱佳彬[2]　李晓勇[1]

（1. 中国商飞上海飞机设计研究院，上海，201210）

（2. 海南航空公司，北京，102200）

摘要：民用飞机的所有权成本在航空公司运营成本中占了很大的比重，对于民机制造商而言，制造并销售民航机，实现盈利是其最终目的。与此同时，关注到飞机作为非常规商品，政府与金融机构也在飞机交易市场扮演着重要的作用。因此，各方对于飞机的价值评估均密切关注，以期实现市场的均衡，促进行业的健康发展。本文针对影响民用飞机价值的因素进行了广泛而全面的分析，在此基础上探讨了价值分析的方法和模型，并阐述了应用方向。本文可以对民机贸易的利益相关方拓展业务形成一定的参考意义。

关键词：民用飞机；价值评估；市场价值；基本价值

Study on Civil Aircraft's Value Evaluation

Zhang Kang[1]　　Zhu Jiabin[2]　　Li Xiaoyong[1]

（1. Market Research Center，SADRI，Shanghai 201210，China）

（2. Hainan Airlines，Beijing 102200，China）

Abstract：Generally speaking，the civil aircraft's ownership in airlines operating costs are accounted for a large proportion. Meanwhile, civil aircraft manufacturer's purpose is manufacturing and sales of aircraft，and taking profits. The government and financial institutions are also important in the market，because the plane is a special goods. So they are all paying close attention to aircraft's value evaluation，finally they realize the balance of the market，and promote the healthy development of the industry. This paper analyzes the influence factors of the value evaluation，and discusses the value analysis method and mode. At last，we points out the application field. For the commercial aircraft trade related party，this paper can be used to guide some business.

Key words：Civil Aircraft；Value Evaluation；Market Value；Base Value

1　相关背景

1.1　基本概念

飞机的价值确定是残值评估的重点，一般使用较多的是飞机市场价值（Market

Value)和基本价值(Base Value)。这两类价值定义又可以各自进一步确定为全寿命期价值(Full Life Value)和半寿命期价值(Half Life Value)。

飞机的市场价值是指目前市场中飞机的"公平价值"。确定市场价值需要在以下前提下进行：

- 特定时间点(如今天)的现货交易价格；
- 单机销售——不享受规模折扣；
- 上述时间点在开放市场条件下交易；
- 两个独立实体间的公开交易；
- 不附加租赁条款——仅限飞机本身；
- 拥有合理的营销时间(12个月)；
- 买卖双方存在交易意愿。

由于飞机市场价值随行业周期而波动，针对其历史数据的研究，得到的成果规律性不够，对于未来期间内飞机市场价值的可预见性也较差，因此提出了飞机基本价值的说法，从而去除了行业周期因素。

总的来说，飞机基本价值是指公平(平衡)市场中的飞机价值。对其的研究具有以下特点：

- 基本价值是指"飞机长期内的基本经济价值"；
- 平衡市场条件(供需平衡)下交易；
- 随时间推移而贬值；
- 将飞机历史价值走势和未来价值预期走势考虑在内；
- 所有未来价值预测均是根据基本价值情境所作出的。

另外一个值得区分的概念是全寿命期价值和半寿命期价值。"半生命周期"(Half Life)是一个标准评估工业的术语，它假设飞机处于机身，发动机，起落架和所有主要部件的使用情况处于全面检修状态，而任何有使用寿命限制的部件都处于使用到一半的状态。

"全生命周期"(Full Life)的价值定义假设在如下维修状态下：机身刚经过重大检查(通常称为D检查)，发动机刚从性能恢复工厂出厂，发动机所有具有使用寿命的部件都未使用过，起落架刚经过大修，所有其他部件参数都定义为处于"半生命周期"(例如维修状况无变化)。

1.2　基本价值和市场价值的对比

正是由于飞机市场价值考虑了行业周期波动的影响，因此其与飞机基本价值的差异是显而易见的(图1)。

下图是飞机市场价值与基本价值关系的理论示意图，下面我们来看看国外两个真实飞机型号的价值数据(图2)，在该图中包含了两方面信息：一是基于研究理论预测的飞机基本价值数据，二是飞机市场价值的历史数据，以此相互比照。

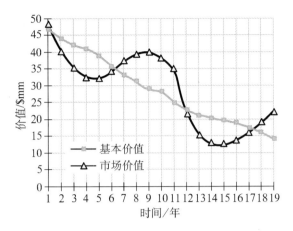

图 1　飞机市场价值与基本价值的理论走势对比

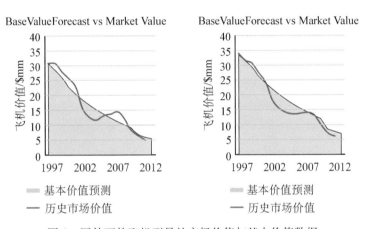

图 2　国外两款飞机型号的市场价值与基本价值数据

2　影响飞机价值的因素

飞机价值的影响因素较为复杂,不仅与国家、行业的宏观经济因素有关,而且与特定机型乃至某架飞机使用情况密切相关。图 3 针对主要因素进行了分类。进一步研究可以发现,这些因素所包括的许多要素共同发挥着作用。

1)宏观经济形势

诸如经济增长率(图 4)、原油价格指数(图 5)、通货膨胀率、贷款利率、全球贸易流、全球旅游流等(后两者体现为旅客和货物运输市场行情)。

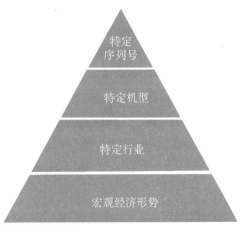

图 3　影响飞机价值的主要因素

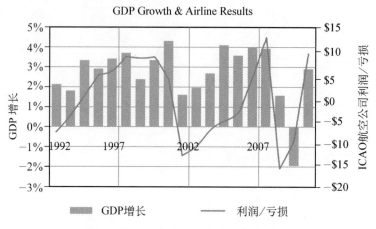

图 4　GDP 是推动航空业发展的主动力

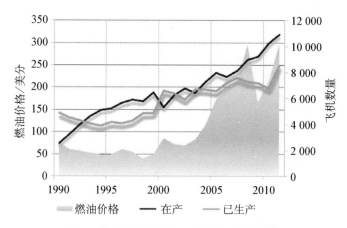

图 5　燃油价格是影响飞机价值的长期因素

2）行业状况

这里的行业状况主要是指民航运输业的经营态势，但是必须留意到飞机制造商的经营决策和生产能力也发挥着较大的作用。包括的要素有民航市场运力状况、储存机队规模、航空公司盈利状况、订单/交付数、监管法规的变化、机队集中更新、制造商稳定性等。

3）特定机型

上述两条大背景决定了飞机交易市场的活跃程度，但不同机型的飞机价值差异仍然很大，这主要决定于该机型飞机机队分布情况、备选发动机受欢迎程度、产品支持和保障能力、生产周期、新机价格、飞机运营经济性、系列化机型的通用性程度。

4）单架飞机状态

影响因素有机龄、重量参数、发动机推力、内饰和设备选项、机体状况、维修状况、累积飞行小时、循环数等。

3 飞机价值评估

3.1 航空业的周期性特点

航空业是典型的周期性行业,乘客旅行需求与 GDP 增长密切相关(图 6),航空公司的盈利因而随世界经济周期变化而上下波动。航空公司订购飞机的能力取决于其资产负债状况,因此航空公司会在高盈利时期订购飞机,盈利高峰期订购的飞机会在 2～3 年后交付,而飞机交付时,经济形势已经开始下滑,经济不景气时航空公司的信用评级和资产负债状况通常较差,值得投资的航空公司屈指可数,航空公司无法融资,导致下一个盈利周期来临时运力不足(见图 7 和图 8)。

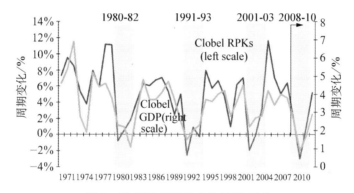

图 6 受 GDP 周期拉动的 RPK 增长

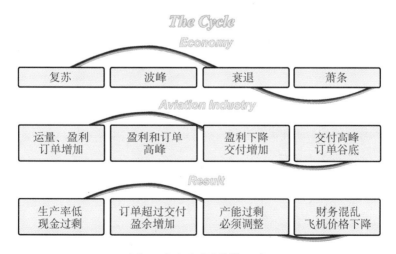

图 7 航空业的顺周期现象

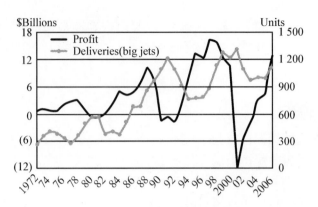

图 8　飞机交付相对盈利的滞后

　　以上因素导致的结果是航空公司总是在经济黄金期订购飞机,而在经济低迷期接收飞机。从图 9 中可以清晰看出航空公司的业绩深受 GDP 增长状况影响,相对较小的 GDP 下滑会引发航空需求的大幅下降;并且,航空公司的盈利会早于 GDP下滑而出现恶化,其盈利状况的好转又会迟于整体经济的复苏。

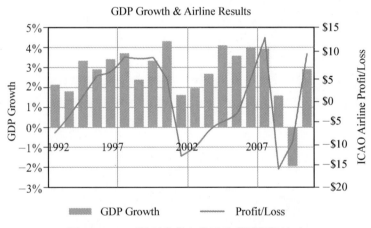

图 9　GDP 下滑引发航空公司盈利周期性波动

3.2　民航机的价值曲线

1) 基本价值和市场价值曲线

　　与所有的机器设备类似,机龄是民航机价值的最大决定因素。在关注民航机的价值曲线之前,首先让我们来关注世界范围内民航机的使用寿命分布情况。(图 10)图中可以看出,双通道飞机的使用寿命相对比较集中。

　　下面以市场上常见机型为例,来观察其基本价值和市场价值。如图 11,不同机型的基本价值随机龄增长一般单调下降(折旧),并且初始价值较高的宽体机贬值更

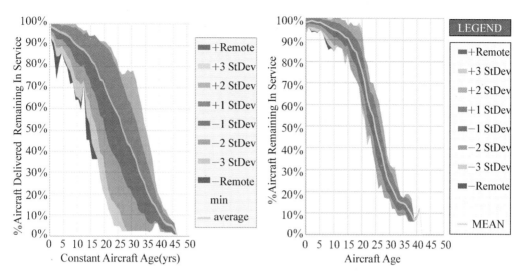

图 10　商用飞机使用寿命（左侧为单通道客机，右侧为双通道客机）

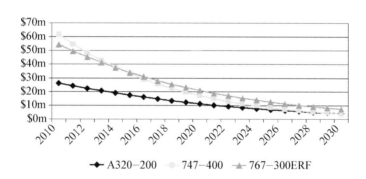

图 11　机型 A320 - 200、B747 - 400、B767 - 300ERF 基本价值曲线

快。与飞机的基本价值不同，其市场价值虽然在长期看来，也是逐渐贬值的，但是在某一个时间段，由于受到经济周期、运输环境等各种因素影响，走势有一定起伏（图 12）。在市场行情较好时，航空公司运力的缺口可能会在二手飞机交易市场带来一波看涨的行情，这一点从飞机的市场价值与基本价值的比值（MV/BV）与飞机交易量的关系可以看出端倪（图 13）。

需要补充的是，在评估行业，飞机基本价值每年均需要根据 GDP、原油价格等经济指标对其进行微调，当出现以下情况时甚至还需要实施重大调整：

■ 市场价值持续高于或低于基本价值预期；

■ 行业基本面出现变化（如飞机制造商破产等）。

2）机型对飞机价值曲线的影响

如上节图 11 所示，在二手飞机市场，我们看到单通道客机在价值表现上要好于

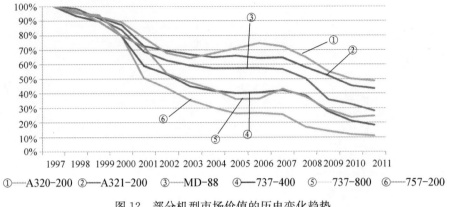

①——A320-200 ②——A321-200 ③——MD-88 ④——737-400 ⑤　　737-800 ⑥——757-200

图12　部分机型市场价值的历史变化趋势

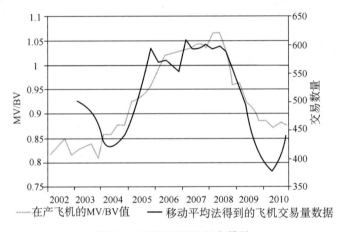

——在产飞机的MV/BV值　——移动平均法得到的飞机交易量数据

图13　MV/BV 和飞机交易量

双通道飞机。这主要归功于单通道飞机拥有更为灵活的市场适应性,并且其较少的选项为飞机在不同承运人之间流转减少了很多麻烦。

表1　单通道客机与双通道客机要素对比

比较要素	单通道客机	双通道客机
平均使用寿命	更长,达 24～28 年	22～24 年
技术更新换代时间	每 14～16 年推出一款新机型	每 8 至 10 年推出一款新机型
机型数量	2 大机型、8 个主要衍生机型	8 大机型、30 个主要衍生机型
机型/发动机组合	12 个衍生机型/发动机组合	50 多个衍生机型/发动机组合
市场需求和飞机产量	需求大,批量生产	小批量,定制化程度高
客户态度	更易为三级运营商采用客户基础好,受欢迎程度高	较小的客户基础,受欢迎程度稍低
价值表现	提高残值潜力	残值潜力受到冲击

另外,研究发现,当后继机型出现时,会对现有二手飞机交易市场的飞机价格形成明显冲击,具体表现如图 14。

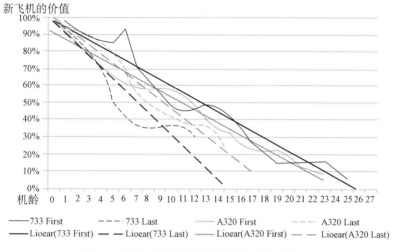

图 14　后继机型出现对现有机型的影响示意

3）维修状况对飞机价值的影响

维修费用在航空公司的日常开支中占有很大比重,因而飞机的维修状况对飞机价值的影响也是显著的(见图 15),一架即将进行大修的飞机和刚刚大修完毕出厂的飞机,其价值往往不具有可比性,尤其对于机龄较大的飞机,尤为如此。

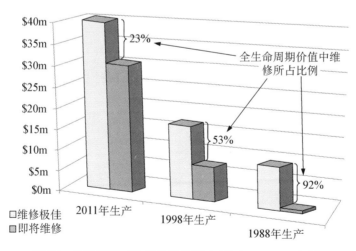

图 15　维修状况对飞机价值的影响(以 A320 - 200 为例)

4）宏观经济因素对于飞机价值的影响

一般而言,GDP 增速放缓或下滑阶段,航空公司以削减运力应对行业衰退,飞机价值相应走低,飞机价值的回升要比 GDP 复苏滞后 2 年。经济复苏导致燃油需求及价格走高,燃油价格持高不下,燃油效率更高的现产机型更受市场青睐。

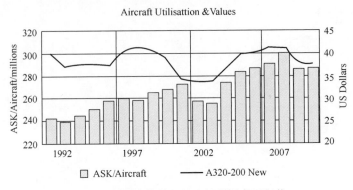

图 16　飞机价值随现役飞机利用率而起伏

4　飞机价值评估方法和模型

当前飞机价值评估主要采用传统方法和计量经济学模型法相结合的方法来评估飞机的价值。

1) 传统方法

传统方法采用一种基于历史交易趋势的模型,该模型按市场潜力将飞机划分成不同的等级,预测的飞机未来价值会依据飞机特性(及生命期)在所属等级范围内移动,需要注意的是基本价值的等级选择不是一成不变的(见图 17)。

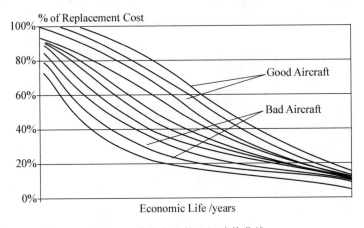

图 17　传统方法的飞机残值曲线

2) 计量经济模型法

采用计量经济模型方法来评估,需要大量的样本数据(图 18),但是能够比较完备的考虑飞机机龄、飞机特性、客户选项(从影响显著的发动机选项到是否加装翼梢小翼)(图19)乃至经济周期对于飞机价值的影响。对于尚未投入市场或者刚刚投入市场的机型,由于缺乏实际交易信息可以采用趋势外推法进行评估(图20)。

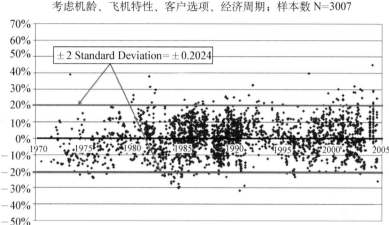

图18 商用飞机残值统计分布(样本数 3 007 个)

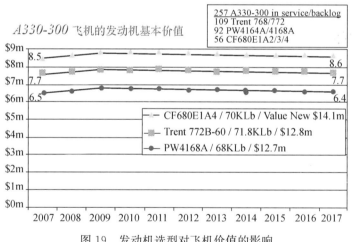

图19 发动机选型对飞机价值的影响

5 发动机价值评估

5.1 航空发动机市场概览

在分析航空发动机的价值之前,让我们首先来研究其市场分布情况。目前全球主要的航空发动机制造商有通用电气(GE)、普惠公司(Pratt & Whitney)和罗罗公司(Rolls-Royce)三大发动机制造商,以及这些公司与其他公司合资组建的 CFM、IAE(International Aero Engines)、EA(Engine Alliance)等公司。部分公司的图标见图 21。

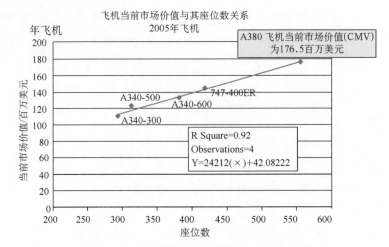

图 20　新型号商用飞机价值评估方法示例

图 21　当前市场主要飞机发动机制造商

　　对以上航空发动机制造商的市场份额的研究发现,无论从发动机的数量还是从发动机的价值指标来衡量,CFM 国际都占据了最大的市场份额,其余各公司的市场占有情况见图 22。图中左边饼图是按发动机数量的统计结果,右边饼图是按发动机价值的统计结果,合计的市场份额为 47 800 台发动机,市场价值计 2 180 亿美元。

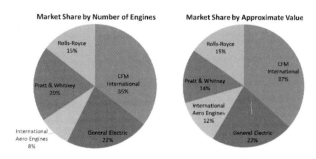

图22　主要飞机发动机制造商的市场份额

5.2　发动机价值分析及其特点

1）影响发动机价值的因素

影响发动机价值的因素比较多,与发动机整个型号的价格、设计寿命、维修性好坏、市场口碑关系紧密,同时也和单台发动机的使用情况有关,并且该型号发动机在生产周期中所处的不同阶段会对市场价值预期产生积极或者消极的影响。归纳起来包括以下几点:

- 价目表价格、既往交易价格;
- 时寿件成本、大修成本(典型数据见表2);
- 产品寿命预期;
- 类似型号的历史表现;
- 在型号生产周期中所处的阶段;
- 单台发动机运营状况;
- 单台发动机已知维修状况。

表2　大修及时寿件(LLP)常见成本

	窄体机发动机(如 CFM56，V2500)	小型宽体机发动机(如 CF6，PW4060)	大型宽体机发动机(如 GE90，Trent 800)
大修费用	\$1.8m～\$2.2m	\$2.5m～\$4.5m	\$4.5～\$5.5m
LLP 备件成本	\$1.8m～\$2.3m	\$3.5m～\$5.3m	\$5.5m～\$6.2m

2）发动机价值分析

为了分析发动机的价值,我们首先来观察其价值构成(图23),从图中可以看出发动机的维修状况对其价值的影响,全寿命周期状态(Full Life)、半寿命周期状态(Full Life)以及即将维修状态(Run Out)的发动机,其价值差异很大,差异主要在于大修成本和时寿件的成本费用。单台发动机在服役周期中的价值示意图见图24。

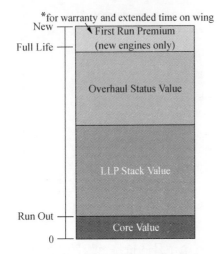

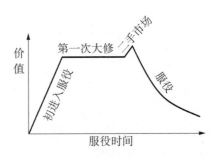

| 图 23　发动机的价值构成示意图 | 图 24　生命周期内发动机价值的走势 |

3) 发动机价值和飞机价值的差异

正如在之前章节所描述的,机龄对于飞机价值的影响是非常巨大的,但发动机不同于此(图 25)。图中示意了两台 GE90 - 115B 发动机与 2007 年产波音 777 -300ER 飞机的价值对比(均处于半生命周期)。假定基本价值是在通胀率假定为2.5%的基础上得出的,单位为百万美元。可以看出线条 1 标注的飞机价值下降明显,而线条 2 标注的发动机在维护良好的情况下,明显走出与飞机不同的价值走势。表 3 将这些差异进行了列举,可以看出发动机相比飞机具有良好的保值特性。

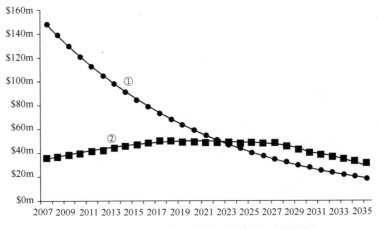

图 25　机龄对于飞机和发动机价值的不同影响

表 3　发动机价值与飞机价值的差异

飞机价值特点	发动机价值特点
生产年份是影响飞机价值的重要因素	不受生产年份影响
维修状况也是影响飞机价值的重要因素	维修状况可占发动机 90% 以上的价值
飞机出厂后，其价值随时间推移而下降	受价目表价格和时寿件价格上浮的推动，发动机价值随时间推移可能出现上升
价目表价格对二手机价值的影响微乎其微	发动机价值(现产)与制造商价目表价格密切相关

值得一提的是，类似于飞机，一款新型发动机投入市场对于现有发动机型号的影响是存在的，但是没有新型号飞机投入市场产生的影响那么迅速。一般而言，新型发动机对现产发动机价值的影响是逐步显现的，只有在新型发动机和新机型数量达到相当规模后才会对现有发动机的价值产生明显影响。图 26 显示的是两款未来将要投入市场的竞争力较好的发动机。

图 26　未来发动机项目

(左：齿轮传动涡扇发动机(P&W)、右：LEAP (CFM))

5.3　发动机价值应用

由于当前民航运输市场相当规模的发动机是采用租赁形式存在的，对于发动机的产权所有人来说，为了保证资产价值，采取了一系列措施，诸如：建立维修储备金制度(保持"全寿命"价值)、与制造商签订服务协议、保存发动机使用过程中的技术记录、建立 PMA 零部件使用政策、实施发动机资产监控等。

对于发动机的实际使用人——航空公司来说，为了延长发动机的使用寿命，往往科学规划、将发动机在机队的不同飞机间调换使用，得到延长发动机寿命的目的。

6　飞机残值应用

飞机价值和残值应用往往和飞机贸易有关，因此利益相关方涉及较广，包括飞机制造商、航空公司、租赁公司、银行机构乃至政府管理部门等。

1）飞机制造商与飞机残值

对于飞机制造商而言，有可能需要和买家签订残值保障协议，在明确了飞机的使用维护要求后，就飞机未来的残值实现给予保障。更进一步，可能会约定对于飞机进行以旧换新或若干年后回收,也就是所谓的飞机回购。为了提升自身飞机产品的残值,飞机制造商需要对已经出售的商品提供技术、航材等服务保障,实现与航空公司/租赁公司的双赢。对于飞机制造商来说,还需要考虑新产品开发中是否需要采用更为先进的技术,运输行业的衰退往往会加速老旧机型的贬值。

2)航空公司与飞机残值

对于航空公司而言,早在进行机队规划和飞机引进时,就需要评估投资风险,考虑折旧残值设定。在飞机运营过程中,飞机保险中的机身险测算一般按飞机市场价值计算。在运营末期,需要考虑飞机账面价值与市场价值的差异,也就是飞机残值的实现过程。另外,航空公司在引进一款新机时,还需要考虑其他航空公司是否采用相同的机型?这类分析应考虑到飞机的技术规格、一般用途等因素,并且要考虑将非标准机型改装为标准机型的成本。

3)租赁公司与飞机残值

租赁公司将飞机视为一种投资,而不是运输工具,因此投资项目在期末所具有的残值是其非常关键的指标。租赁公司在给航空公司提供服务时,需要考虑该型飞机是否是这家航空公司机队的核心机型,因为航空公司通常不会拖欠订购核心机型的债务。同时,还需要考虑这家航空公司的机队规划如何?今天的核心机型明天可能会被其他型号的飞机取代(例如,767被787所替代),需要具备一定的远见卓识。

4)银行与飞机残值

商业贷款是飞机引进最常用的贷款方式。商业贷款中航空公司预付飞机价值的20%~30%,余款由银行贷款支付。银行的借贷风险在于航空公司是否具有在贷款期限内分期还贷的能力。在贷款期内及还款之后,航空公司只要履行还款义务,它都拥有对该飞机的所有权。如有良好资产作担保,银行所承担的风险也相对较低。合同期最长可定为15年,但通常为12年。可采用固定利率或浮动利率。

另一种重要形式是有政府背景的出口信贷,与商业贷款相似,但有出口信贷机构(ECA)向银行作担保,从而获得较低的贷款利率。如果航空公司未能履行还贷义务,作为担保方的ECA将为之偿还贷款,之后再处理与航空公司之间的关系。ECA融资风险较低,因为银行认为一旦航空公司出现问题,政府将为其偿还贷款,航空公司需向ECA支付一定金额以获得其担保。金额大小取决于航空公司本身。

参 考 文 献

[1] 陈金星. 论融资租赁飞机残值的处理[J]. 民航管理,1994(11).

[2] 许海靖. 飞机融资租赁的方式比较与风险研究[D],南京航空航天大学,2002.

[3] 张薇. 国内航空公司飞机融资问题研究[D]. 西南财经大学,2003.

[4] Gibson W E and Morrell. Theory and Practice in Aircraft Financial Evaluation [J]. Journal of Air Transport Management. No. 10 (2004), pp. 427 - 423.

[5] William E Gibson. Airline Fleet Planning and Aircraft Investment Valuation. Air Business

Academy, Toulouse, France.

[6] Benedikt A Mohr. Modeling the Value of Subsystem Flexibility for Aircraft Financial Evaluation using Real Options [F]. 7th Annual Conference on Systems Engineering Research 2009 (CSER 2009).

基于模型驱动的民用飞机数字化设计方法

张莘艾[1]　李浩敏[1]　吴建华[1]

（1. 上海飞机设计研究院总体气动设计研究部，上海，201210）

摘要：飞机设计工作对民用飞机经济性具有决定性的影响。本文概述了基于模型驱动的系统工程方法及典型流程，重点介绍了基于模型驱动的 Rational Harmony 系统工程开发流程。为了提高飞机设计质量、减少设计迭代、降低研制费用进而提高飞机的经济性，采用 Rational Rhapsody 开发工具开展了飞机的需求分析、功能分析和设计综合工作，建立了飞机的用例图、活动图、顺序图和状态机，获得了飞机的架构、功能逻辑、接口等设计结果，并通过数字化方法对设计结果进行了验证。研究表明，基于模型驱动的系统工程方法满足民用飞机设计工作的使用要求，有助于提高飞机设计质量、减少设计迭代、降低研制费用。

关键词：民用飞机；经济性；全寿命周期成本；基于模型驱动的系统工程；设计成本

A Model Based Systems Engineering Design Approach for Civil Aircraft

Zhang Xinai[1]　Li Haomin[1]　Wu Jianhua[1]

（1. General Configuration and Aerodynamics Department，Shanghai Aircraft Design and Research Institute，Shanghai 201210，China）

Abstract：Aircraft design has crucial effect on its economic performance. The model based systems engineering and several typical approaches including Rational Harmony process are introduced. In order to improve the design quality, decrease the design iteration and reduce the design cost, the requirement analysis, function analysis and design synthesis of a civil aircraft are conducted with Rational Rhapsody. The activity diagram, sequence diagram and statechart diagram together with the aircraft architecture, functional logic and interface are obtained. The design results were verified with the executable model of the aircraft. The results show that the model based systems engineering approach is suitable for civil aircraft design and helpful to improve the aircraft design quality, decrease the design iteration and reduce the design cost.

Key words：Civil Aircraft; Economics; Life Cycle Cost; Model Based Systems Engineering; Design Cost

1 引言

随着航空技术的快速发展、市场需求的不断提高,航空公司对民用客机的安全性、经济性、舒适性和环保性都提出了更高的要求[1]。在竞争日渐激烈的市场环境下,有效控制并不断降低飞机的全寿命周期成本(Life Cycle Cost, LCC),已经成为飞机制造商保证所研制的民用客机具有良好市场竞争力的必然选择。研究表明,民用客机全寿命周期成本的85%在设计阶段就已经确定。因此,在飞机设计伊始即考虑飞机的全寿命周期成本、提高飞机设计质量、降低飞机研制费用,对提高飞机的经济性具有重要意义[2][3]。

本文简要介绍了民用飞机全寿命周期成本以及飞机设计对于全寿命周期成本的影响。为了进一步提高飞机设计质量、降低飞机研制费用和全寿命周期成本、提高飞机经济性,介绍了基于模型驱动的数字化系统工程设计方法,并采用该方法开展飞机的需求分析、功能分析和设计综合工作,完成飞机的功能架构、功能逻辑、接口定义等设计工作,并通过运行所建立的飞机可执行模型对设计结果进行验证。将基于模型驱动的数字化系统工程设计方法应用于飞机设计工作,可有效提高设计的质量与效率、缩短研制周期、降低研制费用,对降低民用飞机全寿命周期成本具有积极的作用。

2 飞机设计对全寿命周期成本的影响

全寿命周期成本的概念自从首次提出后,其应用范围不断扩展,已经被广泛应用于各个领域多种产品的设计工作。在民用飞机领域,全寿命周期成本指在一个飞机项目全寿命周期内各个阶段发生的费用的总和,将全寿命周期成本纳入经济性指标体系用以实现成本估算和控制,是项目和技术方案决策的重要依据、决定民用飞机设计成败的重要因素。民用飞机的全寿命周期大致可以分为可行性论证阶段、方案阶段、工程研制阶段、批生产阶段、使用保障阶段、退役处置阶段等六个阶段。相应地,飞机全寿命周期成本一般包括研制费用、生产采购费用、使用费用和处置费用。其中,飞机的研制费用是用于飞机研究、研制、试验及鉴定阶段所需费用的总和[3][4]。

研究表明,民用飞机的研制费用约占全寿命周期成本的10%,生产采购费用约占30%,使用费用约占60%。研制费用在民用飞机全寿命周期成本中所占的比例不大,但早期飞机方案设计对全寿命周期成本具有决定性的影响,全寿命费用的85%取决研制阶段,65%取决于方案阶段。这表明,在飞机设计阶段的方案设计确定了飞机的基本架构,飞机的全寿命周期成本也随之确定[3]。

民用飞机的研制工作是一项周期长、投资大、高新技术密集、系统集成程度高、协调复杂的系统工程。为了满足日益提高的市场需求,飞机的功能愈趋复杂、系统间信息交换量逐渐增大,也进一步增加了飞机的设计难度[5]。因此,在民用飞机设计工作愈趋复杂的情况下,采用先进的设计方法综合多领域的专业技术和系统最终确定飞机合理的设计方案有助于降低民用飞机的全寿命周期成本。

3 基于模型驱动的系统工程方法

系统工程方法被广泛应用于航空航天等众多领域的复杂系统设计工作。采用系统工程方法研制民用飞机,是将整个飞机作为一个系统,具体工作包括飞机顶层需求的定义、飞机架构的综合设计、飞机需求分配、系统及更低层次需求的定义、各层级需求的确认、各系统的详细设计、各系统的生产、系统和飞机的集成,系统和飞机各个层级的验证,最终获得完整符合设计需求的民用飞机[5]。传统基于文件的系统工程方法在众多领域的复杂产品系统设计中发挥了重要作用,但随着产品的功能愈趋复杂、规模不断扩大、交联耦合程度增强,基于文件的系统工程方法已无法满足复杂系统设计开发的需求,亟须对传统方法进行改进与变革以满足全新的使用要求[5—7]。

随着信息技术的飞速发展,基于模型驱动的系统工程(Model-Based Systems Engineering, MBSE)应运而生。该方法将模型的概念引入传统的系统工程方法,提升了模型在产品需求定义、架构设计、需求分配、需求确认与验证等工作中的作用,已逐步成为工业与学术界的研究重点,在各领域复杂系统的设计开发过程中得到广泛应用[7—12]。在应用与发展过程中,已经建立了多种基于模型驱动的系统工程的方法及流程,包括 IBM 公司的 Harmony-SE、国际系统工程委员会(International Council on Systems Engineering, INCOSE)的 OOSEM(Object-Oriented Systems Engineering Method, OOSEM)、IBM 公司的 RUP SE(Rational Unified Process for Systems Engineering, RUP SE)、Vitech 公司的 Vitech MBSEMethodology、NASA 喷气推进实验室(Jet Propulsion Laboratory, JPL)开发的 State Analysis 等[13]。其中,IBM 公司的 Rational Harmony-SE 应用最为广泛,并且已经开发出相应的软件实现基于模型驱动的系统工程方法,在应用过程中具有较为明显的优势条件。

Harmony 方法最初由嵌入式建模工具的重要供应商 I-Logix 公司开发,该公司于 2006 年被 Telelogic AB 公司所收购,IBM 公司于 2008 年收购了 Telelogic AB 公司,最终成为 IBM 公司基于模型的集成系统和嵌入式软件开发方法。Rational Harmony 流程采用系统工程开发过程常用的 V 型全寿命开发模型,包括 Harmony-SE(Harmony System Engineering)和 Harmony-SW(Harmony Software Engineering)两个紧密联系的子流程。由于采用 OMG UML 和 SysML 建模语言,因而可以保证从系统工程到软件工程的顺利过渡。Rational Harmony 基于模型的开发流程如图 1 所示[13—15]。

Harmony-SE 开发流程包括需求分析、功能分析和设计综合三个子阶段,分别完成飞机需求的定义与完善、需求确认与验证、需求分配、架构设计、接口设计等工

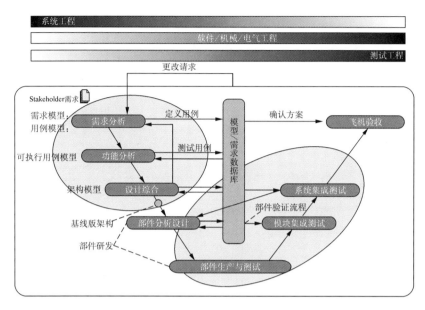

图 1　Rational Harmony 基本流程

作。图 2 给出了需求分析、功能分析和设计综合三个阶段的工作流程图。在需求分析阶段，通过分析市场调研、用户需要以及相关行业规定等内容，归纳提炼所开发产品的涉众需求（Stakeholder Requirement），根据产品涉众需求定义产品需求并建立产品需求与涉众需求之间的链接关系，所定义的系统需求包括系统需要满足的功能性需求以及完成各项功能的效果等的非功能性需求。在此基础上对所定义的系统需求进行分析、归纳，并定义系统用例（Use Case）、建立用例与需求之间的链接关系，以保证设计与需求的一致性与可追溯性。在功能分析阶段，需要对所有的用例进行功能分析，包括定义用例的功能流程、接口定义等工作，通过建立用例的活动图、顺序图和状态机固化设计结果，并通过运行用例可执行功能模型即可确认系统需求并验证设计结果，在确认该阶段的设计工作正确、无歧义的情况下，建立功能与需求间的链接。在功能分析过程中，可以对最初产品需求进行补充和完善，以便进一步提高设计输入的质量。设计综合阶段的工作内容是对产品的架构进行权衡分析，以便在多种可行方案中确定满足所有功能需求和非功能需求的最佳系统架构，并将产品功能分配到已经选定的系统架构，以便对产品进行更进一步的细化设计。在设计综合阶段，可以根据所确定的产品架构、功能分配等工作完成产品接口定义工作。对应于每一个用例，均需要进行需求分析、功能分析和设计综合三个过程工作，三个过程之间存在着紧密联系，通过多次的迭代设计实现完善需求和设计的效果。

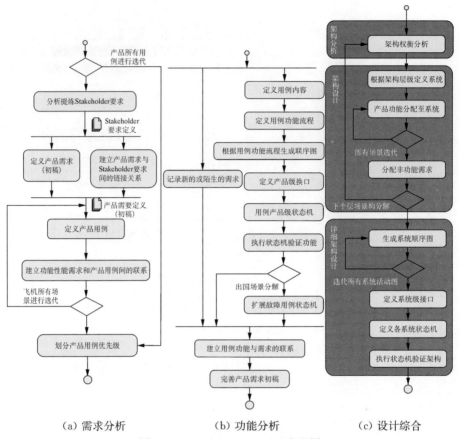

（a）需求分析　　　　　　（b）功能分析　　　　　（c）设计综合

图 2　Rational Harmony SE 流程图

4　设计范例

采用支持 Rational Harmony MBSE 开发流程的 Rational Rhapsody 系统开发工具进行民用飞机设计工作，通过开展需求分析、功能分析和设计综合工作，建立了飞机相关用例的活动图、顺序图和状态机，完成了民用飞机的架构和接口设计等工作。

4.1　飞机需求分析

民用飞机研制过程中首先需要根据市场调研等工作确定利益相关方的需求，在此基础上定义飞机需求作为飞机设计工作的输入。在需求分析阶段，首先将飞机需求导入 Rhapsody 设计工具，以便开展需求分析工作并定义飞机设计用例和用例相关的用户（Actor），通过飞机用例图建立用例与用户之间的关系。如图 3 所示，根据飞机需求并按照飞机工作阶段划分起飞、飞行中和降落三个用例，与之相关的用户包括飞行员、飞行乘组以及 ATC，通过用例图建立了三个 Actor 与起飞、飞行中和降落三个用例之间的关系。在飞机工作过程中，飞机需要根据飞行员、飞行乘组以及 ATC 的具体要求执行既定的功能。

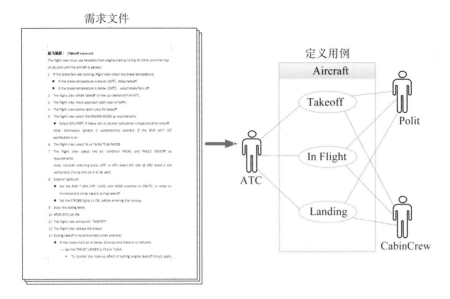

图 3　用例图

4.2　飞机功能分析

Harmony-SE 采用活动图（Activity Diagram）、顺序图（Sequence Diagrams）、状态机（Statechart Diagram）来描述用例的功能和行为。在功能分析阶段，分别建立了飞机的黑盒活动图、顺序图和状态机，如图4、图5和图6所示。

图 4 为飞机起飞用例的黑盒活动图，描述了为满足飞机需求飞机在起飞场景下所需执行的功能及功能间的逻辑关系。例如，飞机需要具备起落架收放功能，在起飞场景下，飞机监测轮载信号和空速信号，当通过监测轮载信号判断飞机已经离开跑道，并且空速满足一定要求后，飞行员即可收起起落架，起落架收进起落架舱后需要锁定起落架，上述过程在黑盒活动图中由监测轮载信号、收起起落架、锁定起落架三个活动来描述，起落架收起前的判据以及飞行员的触发均在模型中进行描述；在飞机起飞阶段，一般不从发动机向空调系统引气，以避免由于引气而造成发动机无法为飞机起飞提供足够的推力，当飞机爬升至一定高度后才允许空调系统从发动机引气，在此过程中飞机需要检查飞行高度。

图 5 左侧为飞机起飞前黑盒活动图，描述了飞机与 ATC、飞行员以及飞行乘组之间的交互关系以及飞机所需进行的功能操作的先后顺序。图 5 右侧是根据左侧活动图生成的黑盒顺序图，顺序图中描述了飞行员、飞行乘组人员和 ATC 与起飞用例之间的交互消息，以及飞机在收到来自 Actor 消息后所执行的具体操作。在起飞前的准备阶段，飞行员需要下载相关数据，在刹车风扇运转的情况下检查刹车温度，如果刹车温度满足使用条件，即可关闭刹车风扇，在进入跑道以前飞行员需要向 ATC 请求批准进入跑道或者起飞，在 ATC 批准后，飞行员需要设定 TCAS、空调、发动机等工作模式，完成后松开刹车准备起飞。

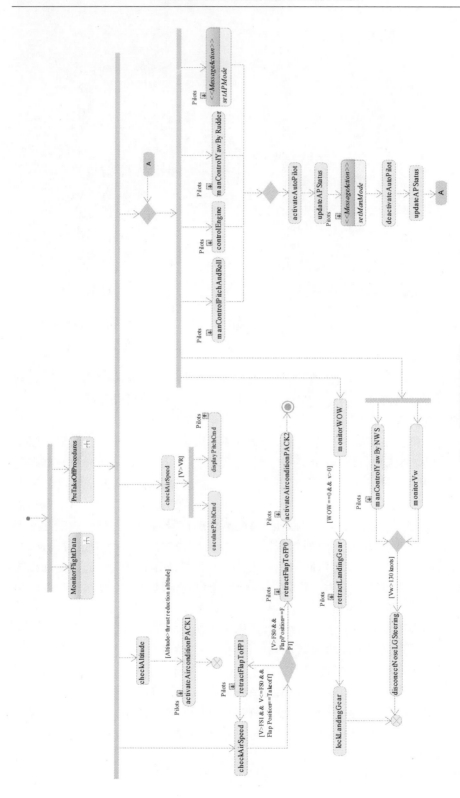

图 4 用例黑盒活动图

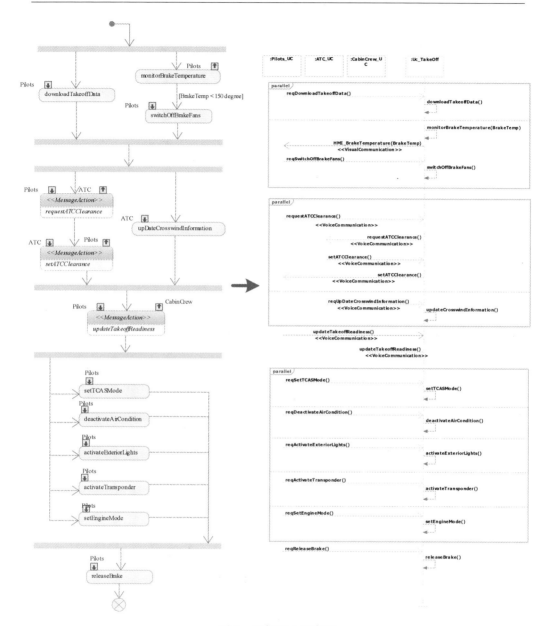

图 5　用例黑盒顺序图

　　用例的状态机图用以描述某一场景下飞机的状态及状态与状态之间的过渡关系。如图 6 所示,在起飞过程中飞机需要通过传感器监测轮载信号,并可通过操纵方向控制飞机的方向,当通过监测轮载信号判断飞机已经离地后,飞机如果接受到来自飞行员的指令后即会收回起落架。所建立的状态机可以分步运行,通过执行该模型可以对功能分析阶段所完成的设计进行初步的验证,如果发现设计结果存在矛盾或者未能满足飞机需求,则可通过改进相关设计结果,进而达到提高设计质量的效果。

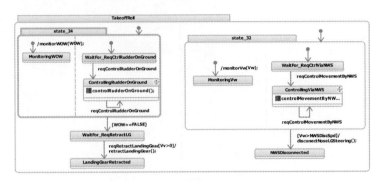

图 6　用例黑盒状态机

4.3　飞机设计综合

在设计综合阶段,首先需要完成飞机的架构设计,即确定为了飞机执行所定义的功能应该包含的系统。在确定了飞机的架构后,再将所有功能分配至相应的系统。图 7 为飞机在起飞场景下的白盒活动图。如图所示,飞机在起飞场景下所需执行的功能被分别分配到航电、起落架、动力装置、环控系统,在功能分配过程中也同时将相应的功能性需求分配至具体的系统,以便各个系统开展进一步的设计工作。为了捕获各个系统在执行既定功能过程中相互之间的交联关系,通过白盒活动图生成飞机用例的白盒顺序图。当分配至两个不同系统的功能之间存在逻辑上的联系时,这时便会使两个系统之间产生接口,如图 8 所示。

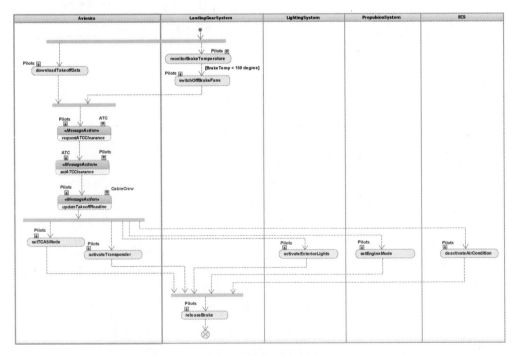

图 7　用例白盒活动图

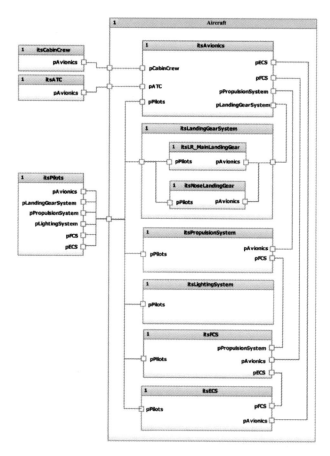

图 8　飞机各系统间接口图

在设计综合阶段,根据飞机白盒活动图和顺序图建立各个系统独立的白盒状态机,通过运行飞机状态机可以实现对设计结果的验证工作,可保证各项设计工作满足所定义的飞机和系统需求,进而避免传统实物或半实物验证方法所固有的周期长、成本高等缺点。

完成飞机级的需求分析、功能分析和设计综合工作后,飞机的需求、功能等均被分配到各个系统,各个系统可根据所需负责的需求与功能进行系统层级的细化设计工作。图 9 为 APU 系统主发引气用例的黑盒活动图,该活动图描述了 APU 起动、接到发动机起动引气信号后引气以及关闭等功能流程,在此基础上可以进行 APU 主发引气用例的功能分析和设计综合工作,具体流程与飞机级相同,此处不再赘述。

5　结论

本文介绍了民用飞机全寿命周期成本的概念以及飞机设计工作对其的重要影响。对基于模型驱动的系统工程方法以及典型流程进行了简单概述,详细介绍了 Rational Harmony 流程及其工作内容。采用 Rational Rhapsody 系统开发工具对飞

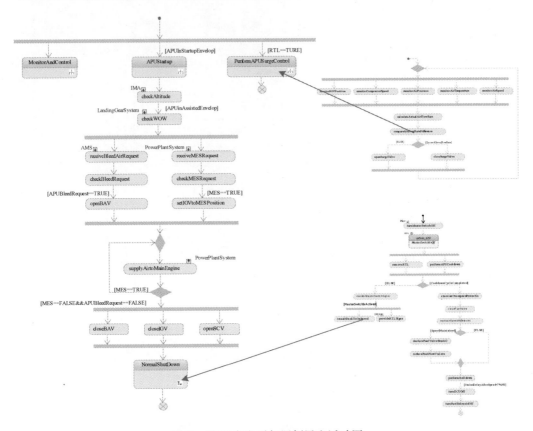

图 9   APU 主发引气用例黑盒活动图

机进行了功能建模,通过开展需求分析、功能分析和设计综合等工作,确定了在既定场景下飞机的架构、功能、功能逻辑、接口等内容,并通过运行可执行模型对设计结果进行数字化验证。研究表明,采用基于模型的系统工程数字化设计方法设计民用飞机,有助于保证设计结果与飞机需求之间的一致性,在设计过程中通过数字化方法验证设计结果可有效提高设计质量,进而减少设计迭代、降低研制费用,对提高民用飞机的经济性具有积极作用。

参 考 文 献

[1]   Raghunathan S, Curran R, Kundu A K, Price M, Benard E. Research into Integrated Aircraft Technologies [R]. AIAA 2003 - 6736,2003.

[2]   彭雄. 浅析影响民用飞机经济性的几点设计因素[M]. //陈迎春. 航空公司运营经济性分析与飞机设计. 上海:上海交通大学出版社,2012.

[3]   飞机设计手册编辑委员会. 飞机设计手册(第 22 册)[M]. 北京:航空工业出版社,2002.

[4]   李晓勇,宋文滨. 民用飞机全寿命周期成本及经济性设计研究[J]. 中国民航大学学报,2012, 30(2):48 - 55.

［5］ Petersen T J, Sutcliffe P L. System Engineering as Applied to the Boeing 777 ［R］. AIAA 92－1010,1992.

［6］ 刘玉生,蒋玉琴,高曙明.模型驱动的复杂系统产品系统设计建模综述［J］.中国机械工程, 2010,21(6):741－749.

［7］ 刘兴华,曹云峰,沈春林.模型驱动的复杂反应式系统顶层设计与验证［J］.系统仿真学报, 2009,21(14):4284－4287.

［8］ 倪忠建,张彦,李漪,等.模型驱动的系统设计方法应用研究［J］.航空电子技术,2011,42(1): 18－43.

［9］ Kul B B, Brain B, Seth M, Wesley E. Integrated approach to architecting, modeling, and simulation of complex space communication networks ［R］. AIAA 2010－1941,2010.

［10］ Mark A F. Developing a heavy lift launch vehicle systems architecture using a modeling based systems engineering approach ［R］. AIAA 2011－7154,2011.

［11］ Oleg V S, Daniel A D, Barrett S C. Conmand, control, communication and information architectural analysis via system-of-systems engineering ［R］. AIAA 2010－8645,2010.

［12］ Michael Fritz, Hans-Peter Roeser, Jens Eickhoff, Simon Reid. Low Cost Control and Simulation Environment for the "Flying Laptop", a University Microsatellite［R］. AIAA 2010－2294,2010.

［13］ Estefan Jeff A. Survey of model-based systems engineering methodologies, Rev. B, INCOSEMBSEInitiative, 23 May 2008.

［14］ Hans-Peter Hoffmann. SysML-based systems engineering using a model-driven development approach. Version 1, IBM Company, 2008.

［15］ Hans-Peter Hoffmann. Systems engineering best practices with the rational workbench for systems and software engineering deskbook. Version 3.1.1, IBM Company, 2010.

飞机全寿命周期成本控制体系研究

邬　斌[1]

（1. 上海飞机设计研究院，上海，200232）

摘要：飞机经济性设计是飞机研制的关键要素，本文试图通过对飞机全寿命周期成本的测算方法、各设计阶段的任务和全寿命周期成本控制方法的体系设计，实现在飞机研制过程中对飞机全寿命周期成本的设计实现和控制，提高飞机研制的产品经济性。

关键词：飞机研制；全寿命周期成本；成本估算；系统工程

Research of aircraft lifecycle cost control system

Wu Bin[1]

（1. Shanghai Aircraft Design Research Institute，Shanghai 200232，China）

Abstract：Economical Design is the key element of aircraft development program，this article identifies the estimation method of aircraft lifecycle cost，the main task of cost control in each design phase and control methods，as a whole system，to realize and get the economical design objectives.

Key words：Aircraft Development；Life Cycle Cost；Cost Estimation；System Engineering

1　引言

随着航空事业的蓬勃发展，自主研发大客飞机已成为转变我国经济增长方式、带动科学技术发展、增强我国综合实力和国际竞争力的重点项目。面对激烈的国际竞争环境，想要在国际市场上占有一席之地，需要为大客飞机制定一个具有竞争力的市场销售价格，并且要回收相应的科研和生产费用，还要获得预想的利润，因此，必须加强对飞机项目的经济性分析与管理，这已成为中国商飞必须面对的关键问题。

中国商飞是一个研发、制造、保障民用飞机并实现商业运作的飞机制造公司。飞机全寿命周期成本的控制对商业运作的飞机设计和制造公司至关重要。同时，飞机研发正处于工程开发阶段，是决定飞机经济性设计指标的关键阶段。因此，目前迫切需要建立飞机全寿命周期的成本控制体系。

飞机全寿命周期费用的成本控制体系，对中国商飞的研发中心来说，是：

（1）基于对飞机方案的经济性指标快速估计；

（2）指标按产品结构的分解下达和集成综合；

（3）多设计方案的优化权衡；

（4）在设计过程中对经济性指标的闭环控制。

因此，对飞机设计中各级别的技术方案的快速经济性指标的估计，是研发阶段的全寿命周期成本控制体系的关键技术，其余三项是飞机设计阶段的组织技术。

2　背景分析和研究

关于飞机全寿命周期费用的成本因素，有以下几种含义：

（1）飞机研发项目的研发费用；

（2）形成飞机研发和制造能力的固定资产基建和工艺设备的投入；

（3）形成产品销售状态的飞机制造成本；

（4）为促进销售、希望设计改善的飞机使用和保障成本。

这几项成本因素构成了飞机的全寿命周期费用/成本。

从全寿命周期费用的成本要素的估计来说，国外航空业采用参数估计模型来估计飞机的全寿命周期成本。目前的参数模型方法，已从二次大战时期的以重量为参数指标，适合大规模批产的飞机成本参数估计关系式，发展到目前注重方案技术特性、以技术复杂度为参数指标的成本参数估计模型。PRICE 软件是其中最具代表性的软件[2]。它能实现对技术方案的快速成本估计。

空客（AIRBUS）和波音（BOEING）已经长期使用这个参数模型。在欧洲超音速客机项目中，空中客车公司使用参数模型软件，在项目早期的设计阶段为决策提供有力的支持。美国波音公司 20 多年来一直使用参数模型成本估算软件，在大型民机系列产品 717、737、747、767、777 的研制过程中，强化对成本的估算与控制工作。

在西方的飞机设计过程中，广泛使用定费用设计的体系[3]，包括建立成本目标、优化成本方案、按费用实施设计活动，以实现在设计过程中对全寿命周期费用的全过程的控制。定费用设计的主要做法是：在方案拟定阶段就应确立按平均单价进行设计的目标；以生产费用和使用保障费用为成本目标，寻求研制费—生产费—使用保障费之间的适当平衡；明确费用增长因素、影响费用的潜在风险和费用—性能折衷方案，实现设计方案的经济性优选；在费用超标时，采取重点纠正措施；强调用新的制造技术来降低费用；费用指标应当按产品分解结构，层层分解、细化，形成定费用设计目标。

目前国内的做法是，在项目设计周期内，决定飞机全寿命周期成本的工程设计部门，没有责任和权利知晓、影响和控制飞机全寿命周期成本。而公司管理、商务和财务部门，通过商务环节，监控飞机的型号经费和制造成本。在立项阶段通过以项目类比法，参照以前项目的历史数据和专家经验，估计出大致的项目资金需要规模。在初步设计阶段，通过主制造商和供应商的联合定义，建立飞机型号初步的成本规模和构成。在详细设计阶段结束,通过招标和工业工程法,建立详细估算。

在国内由于缺少合适的成本测算手段和控制体系,飞机设计部门很难将成本作为飞机设计的重要决定因素,来综合和优化飞机设计方案和细节。例如,在方案阶段,很少对项目的全寿命周期成本进行快速、准确、可靠的估算与分析论证,使全寿命周期成本指标的制定缺少合理的基准;在各备选方案之间,很难从全寿命周期的角度,综合权衡性能、费用、风险等因素,从而确定一个最佳的设计方案;也很难在长达数年的研制过程中,为各种研制成本确定合理的费用指标,并使项目研制过程中的全寿命周期成本的控制缺少合理的基准。这样,决定飞机研制成败的经济性因素,反而没有被设计人员纳入到飞机型号的设计和分析的环节中。

3 飞机全寿命周期成本控制体系的方案建议

为了在飞机研制项目中,将飞机的经济性融入到飞机的研制程序,实现更好的产品经济性。中国商用飞机公司的设计研发中心,目前试图采用参数法和工程法,建立成本测算系统;并引入定费用设计的方法,使飞机设计达到预定的设计目标,实现全过程的设计控制和设计优化。

飞机全寿命周期成本控制体系和方法是:

(1)包含对技术方案的以复杂度为参数的成本参数估算方法和估算手段,实现基于对飞机方案的经济性指标快速估计;

(2)以产品分解结构和工作分解结构为组织手段,实现飞机全寿命周期成本指标按产品结构的分解下达和集成综合;

(3)以构型管理和控制的流程为组织构架,在综合考虑经济性指标的前提下,实现多设计方案的优化权衡和控制;

(4)以工作程序、流程和IT操作平台、设计集成技术等手段,在设计过程中实现成本的经济性指标的闭环控制。

其中,飞机全寿命周期成本控制体系的中心任务是,通过成本估算工具或组织流程方法,实现成本指标的估算。这些工具包括:详细成本估算的工具和组织流程方法;应用估算成本模型的参数估算系统;参数估算系统供应商提供的支持性的历史成本数据库、分析工具和应用管理平台。

在设计阶段,设计研发中心应设立成本管理的专门领导机构;在各个专业培训跨专业的成本工程师,指导具体系统、零部件的成本估算工作;公司或研发中心的商务部门,通过询价来支持成本价格工作;第三方的咨询人员,在一定经费的范围内,提供成本工程的专业支持。

4 飞机全寿命周期成本的测算方法

在飞机全寿命周期成本测算过程[1]中,实际使用的方法包括类比法估算、成本关系式估算、参数法估算、工程法估算。

4.1　项目类比法估算

在规划的早期阶段,项目范围和整体参数的信息量较少的时候使用这个估算方法;通过找相似的飞机项目的成本数据,采用相同的详细程度和条目结构,利用专家的经验,估计飞机全寿命周期各成本因素的指标数额。这个方法的精度在－20％到＋50％。

4.2　成本关系式估算

当设计进展到能计算出飞机的重量、航程、巡航速度、推重比、发动机最大推力等设计指标,应用成本关系式,可以间接地估计出整个飞机在全寿命周期中的各项成本。这个估计方法能有效地估计出各项成本,准确率预计能达到－15％到＋25％。

4.3　参数法估算

在设计图纸完成量在5％～35％时,参数成本估算方法最适合在这个阶段使用,参数成本估计是一个中间层次的估算。典型的估计是依据以技术复杂性的参数指标建立的参数估计模型,使用这种方法的估计全寿命周期成本的估算精度预期在－10％到＋15％之间。

4.4　工程法估算

是一个相对高级别的估算方法,适用于100％发图阶段。估算工作按产品结构,被分解到最底层的产品单元,每一个单元都有一个价格。这些工作单元,是通过工作分解结构和/或产品分解结构组织起来的。这个方法提供了最准确的成本估计,估计精度在±5％区间内。

5　飞机研制各设计阶段划分和成本控制的主要任务

项目的成本预算控制体系可以按设计工作量分为5个阶段。项目设计阶段分15％设计完成、35％设计完成、60％设计完成、100％设计完成、最终设计成本估算。

5.1　设计完成15％阶段

即方案论证设计阶段。根据功能分析,各系统或专业有多个备选方案,基于系统的技术特征,为每一个备选方案估算全寿命周期的各项成本,这样系统方案或初步总体方案就附带了此方案经济性的指标。当经过技术/经济综合权衡后,在确定飞机的方案时,也确定了整个项目的各项基准目标成本。

这个阶段,能决定飞机85％的成本规模。在设计完成15％时,针对飞机产品结构中的每个系统,提交此系统的数量估计和单位成本估计,并沿产品结构树成本,获得基准的全机目标成本。这个阶段,应使用参数估算软件。

5.2　设计完成35％阶段

即初步设计/供应商选定阶段。全寿命周期的各项成本估算,是初步设计的方案权衡方法之一。由于项目经费随可行性研究的结束,已相对固化,纳入预算控制程序。制造成本和使用/维护费用,是定费用设计在这个阶段的主要工作对象。

此阶段交付的成本估计,应基于初步设计确定的结构材料和系统件。在初始阶段,设计状态还未明确结构的材料和用量,同时,飞机定义过程尚未定义各系统的准确构型。这时,可以针对某个结构/系统构型项,建立估算模型,单独估计这类结构/系统构型项的固定价格。随着设计进展、构型定义和供应商的选定,将估计价格逐步替换成实际价格。

在此阶段中期,技术和商务共同工作,对各系统的竞争方案,分别落实各系统项的技术/商务权衡。基于价格的估计或采购实际价格,始终按反映飞机技术方案的主导产品结构进行汇总,不断更新和对比全机的目标成本,从而实现全机的费用性能优选和整体最优。

在初步设计阶段结束时,能确定飞机成本中 95% 的成本规模。这个阶段,应避免使用详细的成本加总估算方式,应使用参数估算软件,实现反映产品技术方案的快速成本估算。

5.3　设计完成 60% 阶段

这个阶段应准备尽可能详细的估算。在这个阶段,尽管设计文件缺少完整的定义,但可在此时获得一个飞机型号主要系统和结构的完整清单。在此清单基础上,基于初步的设计图纸和重要单项的报价,估算出每一个构型项的成本。这个带价格的构型项清单,就完整并有层次的标注出了整个飞机的单机成本。随着项目图纸和规范被进一步发放,将修改和更新成本估计的版本。

在这个阶段,已开始了详细设计,定费用设计是这个阶段的主要成本控制手段。同时,虽然成本预算的规模已经确定,但设计团队可以使用价值工程的方法,分析和确定整个飞机中的重大成本驱动因素,在整个详细设计阶段有针对性的削减飞机的制造成本。这个阶段通过合适的控制手段,如基于飞机产品构型的成本结构,汇报和控制飞机的各项成本。

5.4　设计完成 100% 阶段

图纸和规范 100% 发放时,应准备最终版本的预估算,这个版本是飞机制造商的成本目标,因此受到主要承包商的较大重视。在预估算中应提供每个系统完整和准确的描述,同时文件的所有构型项条目的数量已确定,其价格的报价必须已获得。如果是外包项目,供应商必须按飞机成本结构的要求提交估算文件,文件必须采用书面报价的形式。

5.5　最终版本

估算最终版本的格式要求与最终版本预估算一致,可附加纠正和注解。设计商需要递交包含完整估算文件。当最终版本的工程估算与设计中心的估计最低报价差别 10% 以上,需要进行产品制造和工程的评估分析,以确保承包商不会附加额外成本。通过规范的产品结构,比较估算和供应商的报价,分析整个项目的成本差异。

5.6 变更的测算

合同和规范的变更的成本估计是制造成本的变更谈判的充分依据。这些估计必须采用变更的构型项清单的格式,同时有对照飞机产品结构,将成本分解到材料、工艺过程和人工的最底项的成本细项,便于工程变更内容的谈判和预算控制。设计变更导致的合同价格的增减都要有估算和汇总。

6 建立以预算制为基础的成本控制体系

6.1 目标成本控制体系的要求

飞机的全寿命周期各项成本,可以用目标成本的控制体系的形式加以管理,目标成本的控制体系是管理目标和控制手段的集成。研制、制造和使用维护的成本,都可以作为与现金预算的形式加以管理。目标成本控制体系的具体要求如下:

(1)目标成本的建立,是基于成本估算的。应用成本估算的各类方法,针对产品的全寿命周期内的某一类成本因素,估算整个范围的使用需求,并按产品结构对目标成本范围细分,建立有结构的目标成本体系。

(2)目标成本的使用,是按预算制的管理方法得到控制和实施的。在限额范围内,设计方案和内容可以得到批准,获得预算,继续实施。如果超出目标成本的控制范围,就需要调整方案或目标成本。

(3)建立和更新目标成本的估算工具,有具体的选用要求。估算工具通过相对产品范围建立估算,提供文档来满足目标成本决策者和项目团队的要求。规划和设计的成本估算精度是很重要的[4],因为飞机成本的规模是型号项目是否包括在飞机制造商的产品战略规划中的决定性因素。

(4)目标成本控制体系,是以文件作为建立目标成本、调整目标成本的依据。从项目一开始就要按照项目要求,建立全寿命周期成本估算文件,实施目标成本控制体系。飞机型号项目目标成本的调整,可以作为方案的优选、供应商的选择和项目目标成本范围的更改,通过目标成本控制体系加以评估和控制,进而调整目标成本。

(5)目标成本控制体系,是以产品分解结构为基础,来确定估算方案和估算范围的。成本估算文件,描述按产品分解结构组织的全寿命周期成本。它只描述范围和成本,成本估算范围明确了,就能降低项目范围漏项导致的项目目标成本低估的风险,这样可以建立一个高质量全面的目标成本,作为控制依据。

6.2 成本测算的参数法工具

参数工具和软件程序是建立全寿命周期成本的目标成本的基础。只有通过合适的估算方法和计算,估计出成本目标,加上控制手段,才能建立目标成本控制体系,实施有效控制。成本估算的软件要求是由型号成本管理机构来确定的,或者是由研发中心成本工程师来指定的。

在西方,目前广泛使用 PRICE 参数估计软件,让项目估算师在项目的规划阶段,就能估算和管理项目的成本。PRICE 软件使用预先决定的有关基本飞机参数的模型关系,如重量、设计复杂度、工艺复杂度等参数,来估计详细的工作量。使用的参数模型库来源于过去西方的民用和空军型号项目的模型。同时,项目的预算报告由产品分解结构的形式汇报。

全寿命周期成本的目标成本更改,也必须由建立目标成本基准的同一个基本模型来估计,以确保估算的准确性,将标准构型、特殊成本以及关键构型项的更改反映到目标成本的估算中。在目标成本估算文档中,增加的单项构型项的功能成本,可以作为飞机产品功能/性能的增强,增加到产品的标准目标成本中。

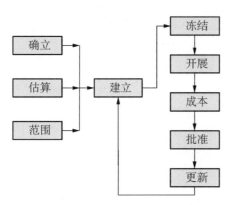

图 1　目标成本计划和控制流程

6.3　目标成本的计划和控制流程

目前,拟通过目标成本的计划和控制流程,把项目的成本估算和项目的过程控制,结合在一起。在飞机设计的各控制阶段,如方案论证、初步设计或详细设计阶段,参照产品结构和成本结构,建立一个飞机项目单一完整的成本分解结构。按设计阶段的次序,逐级打开,形成这一阶段的目标成本控制或定费用设计主要控制用的成本分解结构,见图 1 所示。

把成本因子的金额估计,按结构和成本因子的对应关系,列在一起,形成这一阶段中控制用的带金额成本分解结构。在批准后,成为此阶段的目标成本预算文件。

当发包或设计时,冻结相关范围的预算,在单元设计完成后,确认该单元的实际成本。同时,释放多余预算。项目进展中,成本工程师适时的更新项目预测,预报项目的超支或节省。方案变更时,作出变更估算,在批准后,调整结构和预算。定期的按成本分解结构和成本金额,形成目标成本预算管理报告上报,构成目标成本预算的汇报和控制闭环。

6.4　成本工程的咨询

目前,考虑是否需要成本体系的第三方服务商,维护飞机产品的单位成本指标、规模调整因素、通胀因素、成本质量因素。第三方服务商的飞机成本的指导用于建立符合飞机结构和飞机设计过程实务的成本控制流程,并核对参数估计,或用于增加特殊产品功能的成本基准估算的咨询支持。

7　总结

飞机全寿命周期成本控制体系建立是在研发中心形成飞机设计研制阶段全面控制机制,实现以下功能:

(1) 在项目早期的方案阶段,对项目的研制、生产、使用和保障成本进行估算与分析,开展多种方案基于性能、成本、进度和风险权衡分析,确定最佳的设计方案,制定合理的全寿命周期成本设计目标。

(2) 在方案设计论证阶段,依据很少的数据,使用参数模型软件快速、准确地估算出各种设计条件下的项目成本,项目决策人员直观地看到由于方案的更改对成本的影响作用以及各备选方案的成本,支持对基础设计方案和优化设计的评估和决策分析。

(3) 在设计实施阶段,对各种研制费用和供应商报价的合理性进行评估;为各分系统制定成本目标,跟踪控制项目的成本与进度;使用参数模型软件进行敏感性分析,确定影响飞机项目成本的关键系统和关键因素;进行成本风险分析,在成本估算中考虑风险和不确定因素,对项目成本的风险进行评估和分析,以便尽早采取风险改进措施,降低和控制项目全寿命周期内的风险;在评估供应商的工作中,使用参数模型软件,对供应商报价的合理性进行评估分析,帮助项目管理人员选择最佳的系统供应商,为项目决策提供依据。

(4) 建立成本知识数据库、保证数据的可重用性,对技术进步的趋势进行预测,评估技术进步对成本的影响,及其他多种估算任务。

(5) 建立按方案快速、可靠的估算能力,并把它作为一项飞机设计研究所的战略能力来培养。

因此,飞机全寿命周期成本控制体系的建立能提高飞机设计经济性,对实现民用飞机设计、制造的商业模式,意义重大。

参 考 文 献

[1] (美)防务系统管理学院. 系统工程概论[M]. 2000.
[2] Hardware Lifecycle Estimating Model [CP]. PRICE HL User Manual, 2001.
[3] MIL-766, 美国军用手册, 定费用设计[S]. 1989.8.
[4] (美)NASA. 2008 NASA Cost Estimating Handbook [G]. 2008.

中国机场发展现状、现存问题及应对策略

蒋 政[1]

（1. 上海飞机设计研究院，上海，200232）

摘要：机场作为航空运输和城市的重要基础设施，是综合交通运输体系的重要组成部分。本文对中国大陆地区机场的基本现状、竞争格局、各地区机场发展情况及现存问题进行了较为深入的探讨，并针对现存问题提出了解决方法的建议。

关键词：中国机场；竞争格局；现存问题；应对策略

Current Situations of China Airport and Countermeasures to the Problems

Jiang Zheng[1]

（1. Shanghai Aircraft besign and Research Institute of COMAC，Shanghai，200232）

Abstract：Airport as part of the air transport facilty，is important part of the infrastructure in integrated transport system. The current status，competitiveness，development and issues for civil airport in China is discussed，and some solutions are proposed for the issues currently faced.

Key words：Civil Airport In China；Competitiveness；Issues；Proposals

经过几十年的建设和发展，我国机场体系初具规模，形成了以北京、上海、广州等枢纽机场为中心，其余省会和重点城市机场为骨干，以及众多干、支线机场相配合的基本格局，为保证我国航空运输持续快速健康协调发展，促进经济社会发展和对外开放，以及完善国家综合交通体系等发挥了重要作用，对加强国防建设、增进民族团结、缩小地区差距、促进社会文明也具有重要意义。但机场总量不足、体系结构和功能定位不尽合理等问题仍比较突出，难以满足未来我国经济社会发展需要，特别是提高国家竞争力的要求，进一步优化机场布局和适度增加机场总量已成为未来时期我国机场发展的重要课题。

1 中国机场现状

中国机场经过几十年的建设和发展，初步形成了以北京、上海、广州等枢纽机场为中心，以成都、昆明、重庆、西安、乌鲁木齐、深圳、杭州、武汉、沈阳、大连等省会或重点城市机场为骨干以及其他城市支线机场相配合的基本格局，机场服务能力逐步

提高,现代化程度不断增强,发展速度全球瞩目。

　　截至 2011 年底,我国(不含港澳台地区)共有颁证运输机场 180 个(具体分布见表 1),比 2010 年增加 5 个,2007～2011 五年内共新增 31 个。颁证机场中年内有定期航班通航的机场有 178 个,通航城市 175 个,机场辐射区域覆盖全国 91.5% 的经济总量、77% 的人口和 74% 的县级行政单元。2011 年内定期航班新通航的机场有西藏日喀则、内蒙古阿尔山、内蒙古巴彦淖尔、甘肃金昌、甘肃张掖,迁建完成了库车龟兹机场和揭阳潮汕机场。全国机场密度为平均每 10 万平方千米 1.89 个,各大区机场密度见图 1。

表 1　2011 年各地区运输机场数量

地　区	运输机场数量	占全国比例
东北地区	19	10.6%
东部地区	46	25.6%
西部地区	90	50%
中部地区	25	13.9%

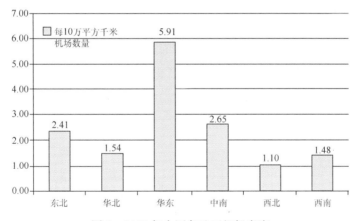

图 1　2011 年全国各地区机场密度

　　2011 年我国机场吞吐量各项指标再创历史新高,其中旅客吞吐量 62 053.7 万人次,同比增长 10%。其中,国内航线完成 57 116.8 万人次,同比增长 9.9%(其中内地至香港、澳门和台湾地区航线为 2003.9 万人次,同比增长 13.6%);国际航线完成 4936.8 万人次,同比增长 10.5%。2002 至 2011 年 10 年间,国内机场旅客吞吐量年均增长率达到 15.6%,除了 2003 年及 2008 年由于受非典和金融危机影响增长率较低外,其余各年份均保持了两位数的高速增长,见图 2。

　　2011 年我国运输机场完成货邮吞吐量 1157.8 万吨,同比增长 2.5%。其中,国内航线完成 750.2 万吨,同比增长 3.9%(其中内地至香港、澳门和台湾地区航线为

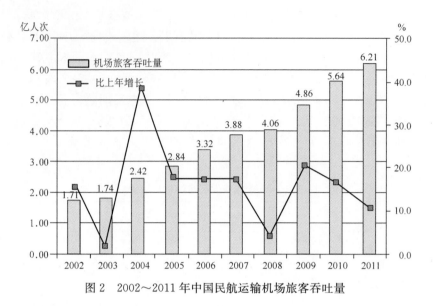

图 2　2002～2011 年中国民航运输机场旅客吞吐量

69.3 万吨,同比增长 0.3%);国际航线完成 407.6 万吨,同比增长 0.1%。2002 至
2011 年的 10 年间货邮吞吐量年均增长率为 13.1%,稍低于旅客运输量的增长速
度。相比而言,货邮吞吐量的增速受 2003 年非典的影响较小,而 2008 年的金融危
机大大减少了贸易往来,特别是国际航线上的货物运输,对增速影响较大,这一负面
影响一直持续到 2009 年,此后 2 年货邮吞吐量增长率连续呈现增长态势,见图 3。

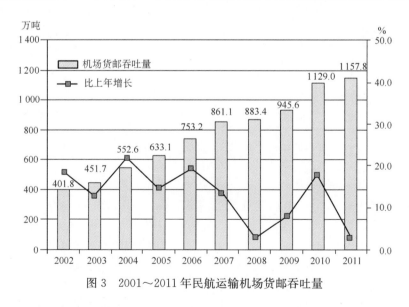

图 3　2001～2011 年民航运输机场货邮吞吐量

　　2011 年我国运输机场完成飞机起降架次 598.0 万架次,同比增长 8.1%。其
中:运输架次为 515.4 万架次,同比增长 6.8%。起降架次中:国内航线 552.8 万架

次,同比增长 8.1%(其中内地至香港、澳门和台湾地区航线为 15.4 万架次,同比增长 13.5%);国际航线 45.1 万架次,同比增长 8.6%。(注:国内、港澳台、国际航线分类按客货流向进行划分)。10 年间机场起降架次年均增长率为 11.9%,总体保持了高速增长,见图 4。

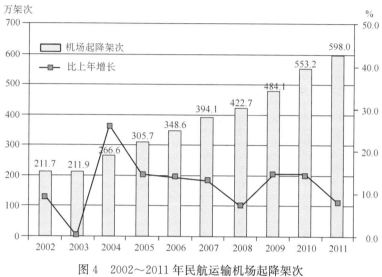

图 4　2002~2011 年民航运输机场起降架次

2011 年全国定期航班通航机场中,年旅客吞吐量在 1000 万人次以上的为 21 个,比上年增加 5 个,完成旅客吞吐量占全部机场旅客吞吐量的 75.1%;年旅客吞吐量在 100 万人次以上的有 53 个,比上年增加 2 个,完成旅客吞吐量占全部机场旅客吞吐量的 95.2%,见表 2。北京、上海和广州三大城市的 4 个机场旅客吞吐量占全部机场旅客吞吐量的 32.0%。2011 年全国各地区旅客吞吐量的分布情况是:华北地区占 17.7%,东北地区占 6.1%,华东地区占 29.8%,中南地区占 24.4%,西南地区占 14.6%,西北地区占 7.5%。

表 2　2011 年旅客吞吐量 100 万人次以上的机场数量

年旅客吞吐量	机场数量	比上年增加	吞吐量占全国比例
1000 万人次以上	21	5	75.1%
100~1000 万人次	32	2	20.1%

2011 年全国定期航班通航机场中,年货邮吞吐量在 10000 吨以上的有 47 个,与上年持平,完成货邮吞吐量占全部机场货邮吞吐量的 98.6%,见表 3。全国各地区货邮吞吐量的分布情况是:华北地区占 17.0%,东北地区占 3.7%,华东地区占 43.8%,中南地区占 22.9%,西南地区占 9.5%,西北地区占 2.1%,新疆地区占 1.0%。其中北京、上海和广州三大城市机场货邮吞吐量占全部机场货邮吞吐量的 54.9%。

表3　2011年货邮吞吐量万吨以上的机场数量

年货邮吞吐量	机场数量	比上年增加	吞吐量占全国比例
10 000 吨以上	47	0	98.6%

　　2011年客运前十位机场业务量均保持正增长。旅客吞吐量排名第一的是北京首都国际机场,其旅客吞吐量为7 867.45万人次,同比增长6.4%,排名继续保持在世界第二位,起降架次53.32万次,远超其他各大机场,仍为中国最繁忙的机场;货邮吞吐量排名第一的是上海浦东国际机场,其货邮吞吐量达到308.53万吨,同比降低4.4%。起降架次34.41万架次,同比增长3.6%。上海虹桥国际机场旅客吞吐量3 311.24万人次,货邮吞吐量45.41万吨,在全国机场中分别排在第四、第六位。上海浦东和虹桥两机场合计吞吐量7 456.02万人次,已接近首都机场。2011年,旅客吞吐量排名前十的机场相比2010年保持不变,排名稍有变化,成都双流机场发展迅猛,一举超越深圳宝安机场跻身前五,杭州萧山机场下降一位至第十位。总体而言,中西部机场发展明显快于东部机场,西部的四大机场旅客吞吐量增长率均在10%以上,东部均在10%以下。

2　竞争格局分析

　　从机场竞争结构看,2011年北京、上海、广州三大城市的四家机场占据了客运市场31.95%的份额,而这一比例在2001年和2006年分别为39.44%和36.48%(见图5),总体而言三大城市机场客运量占比逐步下降,表明随着枢纽机场的日益饱和,客运量有向中小型机场分散的发展趋势。2011年北京首都、上海浦东、广州白云、深圳宝安四大机场占据了货运市场58.16%的份额,近几年占有率基本保持在60%左右,相比2001年的47.50%已大幅提高,主要原因是这四大机场所在城市经济发达,机场货物处理效率高且陆路交通便利,适宜将货物运抵机场后由陆路交通进行转运,从货流运输成本的角度分析,这四大机场的货运占有率仍将保持高位。

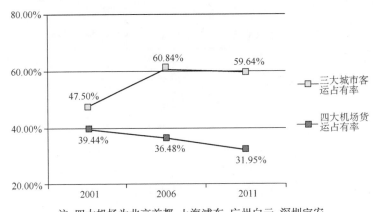

注:四大机场为北京首都、上海浦东、广州白云、深圳宝安

图5　主要机场客运占有率、货运占有率变化

从各大区情况来看,2011 年,西北地区机场增长最为迅速,旅客吞吐量、货邮吞吐量、起降架次三项指标增幅分别为 17.4％、10.0％、17.5％,增幅排名我国六大区首位。在经历了 2010 年井喷式的增长后,2011 年华东地区各项指标增速显著放缓,货邮吞吐量甚至出现了负增长,见表 4。

表 4　2011 年我国各地区机场业务量增长情况

地区	旅客吞吐量增幅		货邮吞吐量增幅		起降架次增幅	
	2011	2010	2011	2010	2011	2010
东北地区	11.7％	15.4％	3.7％	13.6％	12.8％	9.2％
华北地区	9.3％	17.4％	5.1％	8.2％	6.1％	12.5％
华东地区	7.2％	20.5％	−1.2％	24.0％	5.9％	10.8％
中南地区	9.4％	10.7％	4.6％	23.4％	8.3％	7.9％
西北地区	17.4％	26.8％	10.0％	25.8％	17.5％	21.6％
西南地区	13.2％	11.0％	9.0％	11.3％	7.8％	34.9％

旅客方面,全国各地区的旅客吞吐量分布情况是:东北地区占 6.1％,华北地区占17.7％,华东地区占 29.8％,中南地区占24.4％,西北地区占 7.5％,西南地区占14.6％,具体分布见图 6。支线机场较为集中的西部十省市区(四川、重庆、贵州、云南、西藏、陕西、甘肃、青海、宁夏和新疆,下同)机场旅客吞吐量及货邮吞吐量占全国比重为 22.1％,比去年有所上升,见图 7。

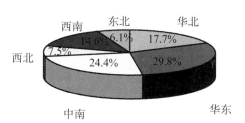

图 6　2011 年国内各地区旅客吞吐量分布

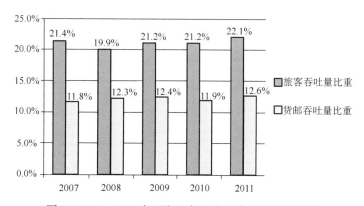

图 7　2007～2011 年西部机场业务量占全国比重走势

　　货物方面,全国各地区货邮吞吐量的分布情况是:东北地区占 3.7%,华北地区占 17.0%,华东地区占 43.8%,中南地区占 22.9%,西北地区占 3.1%,西南地区占 9.5%,具体分布见图 8。西部十省市区机场货邮吞吐量占全国的比重为 12.6%,较上年上升 0.7 个百分点。

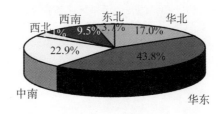

图 8　2011 年国内各地区货邮吞吐量分布

3　全国各地区机场运营情况

3.1　东北地区

　　2011 年,东北地区无新增机场,总数仍为 19 个,平均每 10 万平方公里 2.41 个。年旅客吞吐量 100 万人次以上机场 5 个(大连周水子机场、沈阳桃仙机场、哈尔滨太平机场、长春龙嘉机场和延吉朝阳川机场),其中沈阳机场旅客吞吐量首次突破1000万人次。

　　2011 年东北地区机场的旅客吞吐量、货邮吞吐量、起降架次三项指标分别同比增长 17.71%、3.71%、12.83%。

　　总体而言,东北地区机场业务量(见表 5)在全国的比重仍然较小。对于地势辽阔,冬季地面交通不畅的东北,支线机场是其未来主要发展方向。

表 5　2007～2011 年东北地区机场业务量

年份	旅客吞吐量/人次	货邮吞吐量/吨	起降架次/次
2007	21 768 499	310 386	197 701
2008	24 398 053	332 338	226 817
2009	29 361 497	358 916	267 415
2010	33 873 164	407 642	292 119
2011	37 838 722	422 751	329 597

3.2　华北地区

　　2011 年,华北地区新增机场 2 个(内蒙古阿尔山机场、内蒙古巴彦淖尔机场),总数达到 24 个,平均每 10 万平方千米 1.54 个。年旅客吞吐量 100 万人次以上机场 8 个,仅北京首都机场旅客吞吐量超过1000万人次。

　　2011 年,华北地区机场的三项指标中,旅客吞吐量增幅达到 9.35%,低于全国

10.11%的平均水平。其中北京首都机场、北京南苑机场和天津机场的旅客吞吐量增幅分别为 6.4%，23.6%，3.8%。

2011 年，首都机场再次成为全球旅客吞吐量第二大的机场（见表 6），仅次于美国亚特兰大机场。当前首都机场共有三个航站楼，两条 4E 级跑道，一条 4F 级跑道。2008 年启用的 T3 航站楼为全球最大的单体航站楼，建筑面积 90 多万平方米，新增 99 个机位。作为中国的门户机场，首都机场起降极为频繁，每天要为 70 多家航空公司的 1400 多个航班提供飞机起降服务，连通全球 208 个城市。但是，随着吞吐量逐渐趋于饱和，首都机场的发展也遇到了瓶颈，旅客吞吐量的增幅已连续 5 年低于全国平均水平，这也间接造成了周边南苑机场、天津机场、正定机场的迅猛发展，现在坐飞机到达北京周边机场，再换乘高铁前往北京已为很多人所接受。

表 6　2007～2011 年华北地区机场业务量

年份	旅客吞吐量/人次	货邮吞吐量/吨	起降架次/次
2007	66 157 556	1 608 406	603 900
2008	72 070 106	1 617 022	680 952
2009	85 525 141	1 735 132	743 596
2010	100 371 281	1 876 853	836 543
2011	109 755 511	1 972 878	887 565

近年来，在政府的大力支持下，河北省机场业实现了快速发展。2011 年，河北省全省机场运营航线 80 余条，通航城市 51 个。其中，石家庄正定机场运营航线 67 条，通航城市 44 个，包含 9 条低成本航线。河北省机场管理集团看准自身优势，制定了航空大众化的发展战略，联合河北航空、春秋航空将正定机场打造成低成本航空枢纽，分流过度饱和的首都机场旅客，实现了自身的大跨越发展，5 年间旅客吞吐量年均增长率达到 45%，成为了推动地区经济增长的引擎（见图 9）。伴随着京津冀一体化进程的加快，在政府政策的支持鼓励下，河北省的机场业必将持续快速发展。

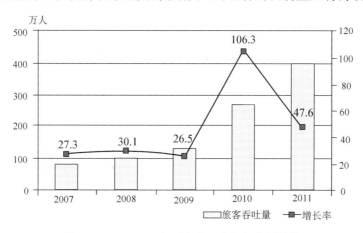

图 9　2007～2011 年石家庄正定机场发展趋势

3.3　华东地区

2011 年,华东地区机场数量不变,总数仍为 39 个,为全国机场数量最多的地区,平均每 10 万平方千米 5.91 个,密度远高于其他各大区。华东地区拥有年旅客吞吐量 100 万以上机场 15 个,其中 1000 万以上机场 6 个(上海浦东机场、上海虹桥机场、杭州萧山机场、厦门高崎机场、南京禄口机场、青岛流亭机场)。

2011 年,华东地区机场的旅游吞吐量和起降架次同比实现增长,增幅分别为7.25% 和 5.86%,货邮吞吐量同比降低 1.27%。三项指标在全国机场中占有的比例很高,旅客吞吐量占全国机场的 29.81%,货邮吞吐量占全国机场的 43.81%,起降架次占全国机场的 27.42%,均排名全国第一(见表 7)。

表 7　2007～2011 年华东地区机场业务量

年份	旅客吞吐量/人次	货邮吞吐量/吨	起降架次/架次
2007	117 618 109	3 946 384	1 155 883
2008	122 753 491	4 092 140	1 231 661
2009	143 211 888	4 141 219	1 398 225
2010	172 507 679	5 136 297	1 548 773
2011	185 009 254	5 072 113	1 639 490

华东地区机场三项指标的增减幅度直接影响我国机场三项指标的增减幅度,而影响华东地区三项指标最关键的因素是上海两大机场三项指标的增幅。相比于2010 年世博会期间的高速发展,2011 年浦东、虹桥两大机场业务量增速明显放缓,旅客吞吐量分别同比增长 2.14%、5.79%,货邮吞吐量同比分别减少 4.63%、5.81%。飞机起降架次同比增长 3.6%、4.96%。浦东机场旅客吞吐量排名全球21,货邮吞吐量保持全球第 3,两大机场的总旅客吞吐量已经接近首都机场,达到7456 万人次。2010 年,虹桥机场启用了新的 T2 航站楼,使机场飞机起降能力提高了 60%,高峰时可满足每小时起降 45 架飞机。机场保障能力也得到大幅提升,达到年旅客吞吐量 4000 万人次,货邮吞吐量 100 万吨,起降 30 万架次。按照规划,两大机场到 2015 年时总旅客吞吐量将达到 1.1 亿人次。

3.4　中南地区

2011 年,中南地区机场数量没有变化,仍为 27 个,平均每 10 万平方千米 2.65个。中南地区拥有年旅客吞吐率 100 万以上机场 12 个,其中 1000 万以上机场 7 个(广州白云机场、深圳宝安机场、长沙黄花机场、武汉天河机场,海口美兰机场,三亚凤凰机场,郑州新郑机场),其中海口美兰机场、三亚凤凰机场和郑州新郑机场是首次突破千万人次大关,海南同时拥有了两个千万级机场。

2011 年,中南地区机场的三项指标中旅客吞吐量、货邮吞吐量和飞机起降架次增幅分别为 9.38%、4.64% 和 8.34%。三项指标在全国机场中占有的比例依然较高,分别占全国总量的 24.35%、22.86%、27.7%,对全国总量的影响程度仅次于华东地区(见表 8)。

表 8　2007~2011 年中南地区机场业务量

年份	旅客吞吐量/人次	货邮吞吐量/吨	起降架次/架次
2007	98 984 587	1 730 562	1 139 860
2008	105 842 349	1 708 359	1 246 314
2009	124 804 399	2 049 478	1 417 751
2010	138 161 311	2 528 787	1 529 059
2011	151 124 973	2 646 122	1 656 539

影响中南地区三项指标最关键的因素是广州白云机场和深圳宝安机场三项指标的增幅,2011 年,白云机场旅客吞吐量的增幅为 9.92%,货邮吞吐量的增幅为 3.1%,飞机起降架次的增幅为 6.09%;宝安机场旅客吞吐量的增幅为 5.74%,货邮吞吐量的增幅为 2.38%,飞机起降架次的增幅为 3.43%。

3.5　西北地区

2011 年,西北地区机场总数为 16 个,新增 2 个(甘肃金昌机场,甘肃张掖机场)。由于西北地域广阔,平均每 10 万平方千米机场数量仅 1.1 个,密度为全国最低。西北地区拥有年旅客吞吐量 100 万以上机场 5 个(西安咸阳机场、乌鲁木齐地窝堡机场、兰州中川机场、银川河东机场、西宁曹家堡机场),其中仅西安机场达到年旅客吞吐量 1000 万以上。

2011 年,西北地区三项指标均实现了较大幅度的增长,其中旅客吞吐量增长 17.37%,货邮吞吐量增长 10.03%,起降架次增长 17.54%(见表 9)。

表 9　2007~2011 年西北地区机场业务量

年份	旅客吞吐量/人次	货邮吞吐量/吨	起降架次/架次
2007	24 461 174	237 601	267 139
2008	24 410 039	238 680	265 872
2009	31 165 774	259 358	317 067
2010	39 515 287	326 236	385 614
2011	46 380 218	358 942	453 260

2011 年,西北地区最大机场西安咸阳机场继续保持快速增长,旅客吞吐量、货邮吞吐量和飞机起降架次三项指标分别增长 17.51%、9.18%、12.56%,除 2008 年受金融危机影响增速放缓外,已连续多年高速增长。乌鲁木齐地窝堡机场连续多年保持强劲增长,2010 年旅客吞吐量增长 21.1%,一举跨入千万级机场行列,成为推动新疆经济增长的强力引擎。

近年来西北地区民航业的快速成长与国家和地方政府众多政策的支持息息相关,未来随着更多扶持政策的出台,西北地区将以西安咸阳机场为核心枢纽,继续重点发展支线航空,西北地区机场业的快速增长势头必将得到延续。

3.6 西南地区

2011年,西南地区新增机场1个(西藏日喀则机场),总数达到37个,数量仅次于华东地区,平均每10万平方公里机场数量1.44个。西南地区拥有年旅客吞吐量100万以上机场8个,其中有3个超过1000万,分别是成都双流机场、昆明巫家坝机场、重庆江北机场。

2011年,西南地区机场业务量增长速度慢于全国,三项指标增幅分别为13.20%、8.96%、7.83%,且起降架次的增长是由于绵阳机场承担大量飞行训练的起降,使机场全年起降次数达到207140架次(见表10)。

表10 2007~2011年西南地区机场业务量

年份	旅客吞吐量/人次	货邮吞吐量/吨	起降架次/架次
2007	58595737	777644	576294
2008	56288066	845051	575126
2009	71994792	911543	696656
2010	79883578	1014057	939608
2011	90427856	1104872	1013213

与西北地区类似,西南地区机场业务量同样集中在几个主要的机场。虽然三项指标在全国总体占比不高,但三大机场却均位列全国10大机场行列。2011年,西南地区三大机场旅客吞吐量均实现了又一次跨越,成都机场突破2900万,昆明机场突破2200万,重庆机场突破1900万。三项指标中,成都机场的增幅分别为12.7%、10.5%、8.2%,昆明机场的旅客吞吐量和起降架次增幅分别为10.3%、5.7%,而货邮吞吐量下降0.4%,重庆机场的增幅分别为20.6%、21.4%、14.5%。

4 现存问题

4.1 机场的总量相对较少,分布不均,航线网络不完善

经过多年的发展建设,我国已经构建了基本的机场网络,辐射范围覆盖了全国四分之三的县级行政区域,但与民用航空已发展较为成熟的西方发达国家相比,仍有较大差距。

美国的国土面积与我国相近,但是其民用机场数量却要远远地超过我国,截至2011年底,我国颁证机场总量为180个,而2011年美国仅有定期航班通航机场就超过800个,是我国的4倍多;从机场地区分布看,我国东部地区每10万平方公里有4

个机场,中部地区为 1.6 个,西部地区为 1 个,东部地区密度大,中西部地区密度小。东部地区机场密度是西部的 4 倍。

4.2　大型国际机场中转乘客少

中转旅客比例是衡量枢纽机场的主要标准,国际大型枢纽,如亚特兰大、阿姆斯特丹、迪拜、香港等,其中转旅客比例往往都大于 30%。全球吞吐量最大的亚特兰大机场,中转率甚至超过 70%。而我国国内的三大枢纽机场,北京首都、上海浦东以及广州白云,旅客中转率均在 10% 以下,离建设成国际大型枢纽机场的战略地位还有较大差距。

4.3　枢纽机场过度拥挤,中小机场亏损严重

当前,国内大中小型机场发展极不平衡。由于空域紧张,资源有限,国内多数大型机场处于超负荷运行状态,而很多中小型机场由于经济环境、地理位置的不利因素,飞机起降架次极少,甚至不能保证每天一班飞机的起降。

2011 年,年旅客吞吐量排名前 15 的机场完成的旅客吞吐量占全部机场旅客吞吐量的 64.88%。而另一方面,全国还有 50 多个机场年旅客吞吐量不足 10 万人,这其中有多达 35 个机场年旅客吞吐量还不足 5 万人,10 个支线机场的旅客年吞吐量甚至不足 1 万人。

中小型机场的业务量过小,导致机场的资源浪费,不利于机场的发展,也不利于当地经济的发展,甚至会成为当地经济的负担,也不利于国家民用航空运输业整体的发展。

4.4　市场化经营程度低,机场的企业化、商业化经营发展不成熟

在航空运量快速上涨的带动下,我国机场的总体收入水平迅速上升,但利润水平及每旅客收入依然较低,与世界其他主要机场存在不小的差距。

我国的航空性收入受政府管制,国内机场的相关航空收费相对其他民航发达国家已处于高位,因此利润水平和每旅客收入差距主要来源于非航空业务收入。

4.5　机场建设盲目性大

由于对机场经济特性认识的局限性,笼统地认为机场是公益性的基础设施,一些航空需求不旺盛的地区也盲目上马机场项目,建成后旅客稀少,亏损严重。机场建设求大求全现象严重,机场建设不考虑收益与成本,最终实际旅客流量远远小于机场设计容量。

5　应对策略

5.1　合理规划,加快建设,优化运输机场布局

在充分考虑政治、经济等要素的条件下,合理推进新机场建设,加强经济发达地区的多机场体系建设,推动枢纽机场与周边机场的功能互补。

对于当前负荷较大的机场,应积极推进机场改扩建工程,提升机场容量。条件

成熟地区可考虑建设新航站楼甚至新机场。

5.2 推进国际枢纽机场建设

北京首都、上海浦东、广州白云三大机场均已将建设成国际一流的大型枢纽机场作为自身的战略目标。为达到这一战略目标,应加强四方面的能力建设：

（1）高效的地面保障；

（2）快捷的中转流程；

（3）合理的航班波设计；

（4）便捷的通关政策。

5.3 发展低成本航空,与中小型机场合作,共同发展

低成本航空的兴起是国际民航业发展到一定阶段的必然趋势,欧美等航空业发达的国家,都有较为成熟的低成本航空体系。我国民航业当前处于快速发展期,航空运量已达到世界第二位,但低成本航空的发展却较为滞后,未来应加强低成本航空的建设,刺激运量需求的进一步增长。

低成本航空的兴起同时可以带动中小型机场的发展,当前处于亏损状态的中小型机场,应加强与低成本航空公司的合作,通过较低的服务收费吸引较多的航班起降,分流日益饱和的大型枢纽机场的航空运量。石家庄正定机场正是因为与春秋航空开展了战略合作,才带来了近年来吞吐量的爆发式增长。

5.4 加快机场市场化运营,提高非航空业务收入比例

机场非航空业务收入是机场经营收入的重要组成部分,欧美等国家机场发展成熟,非航空业务收入往往达到总收入的 70%～80%,部分甚至达到 90%。

我国机场收入中非航空业务收入占比较低,有非常大的上升空间。可参考美国机场业的发展经验,机场所有者仅对机场的运营执行管理职能,不直接参与商业资源的经营,将其交给专业化的企业经营,以提高运营效率,从而刺激非航空业务的发展,提升非航空业务收入比例。

6 总结

中国机场业经过 30 年的发展,已经成为世界机场舞台上一股重要力量,运输保障能力有了质的提升。2011 年国内运输机场已达到 180 个,覆盖全国 91% 的经济总量、76% 的人口和 70% 的县级行政单元。北京首都机场、上海浦东机场、广州白云机场旅客吞吐量分别达到 7867、4144、4504 万人次,首都机场客运跃居全球第二,浦东机场货运全球第三,旅客吞吐量超过 1000 万人次的机场数量达到 21 个。机场的快速发展给旅客出行带来了方便,有力带动了地方经济的发展,形成了一系列围绕机场的空港经济产业,创造了大量就业机会。伴随着中国经济的持续快速增长,在国家政策的支持下,中国的机场业必将迎来更好的发展。

参 考 文 献

［1］ 杨秀云,卓少杰,王新安.中国机场业管制改革的演进与有效性[J].西安交通大学学报,2010(5):29-33.

［2］ 李丽娟.中国机场运营管理模式需要一场变革[J].中国民用航空,2009(106):46-47.

［3］ 陈共荣,刘志.论民航机场对地方经济发展的梯层贡献[J].求索,2008(8):11-13.

［4］ 张国华.大型空港综合交通枢纽规划设计技术体系[J].机场发展论坛,2011,7.

［5］ 我国机场建设"十一五"回顾及"十二五"展望[J].2011,3.

［6］ Marques R C, Brochado A. Airport regulation in Europe: Is there need for a European Observatory? [J]. Transport Policy, 2008,15:163-172.

Aircraft Direct Operating Costs and Cost Index Usage — a Review

Xiaoyong Li[1]   Wenbin Song[2]   Andrew Pfeil[3]   Yingchun Chen[1]

(1. Commercial Aircraft Corporation of China)

(2. School of Aeronautics and Astronautics, Shanghai Jiao Tong University)

(3. Stirling Dynamics, UK)

Abstract: This paper first reviews DOC techniques, in particular the Cost Index (CI) method, and how these techniques are used by airline operators including the different strategies adopted both by traditional and low-cost airlines. There is a top-level overview of the interaction between the Flight Management System (FMS) and Cost Index (CI) and how the FMS uses CI to optimise flight profiles. The paper also covers aspects to be considered by manufacturers during the design stage that reduce fuel consumption and improve aircraft efficiency. Such improvements can result in lower operating costs. Some case studies that discuss work that Stirling has been involved in to improve aircraft performance have also been included. This extended paper is based on a technical report originally prepared by Stirling Dynamics.

Key words: Direct Operating Cost; Cost Index; Flight Management System; Flight Profile

1 Introduction

Aircraft Operating Costs consist of:

- Direct Operating Costs (DOC)
- Indirect Operating Costs (IOC)
- Cash Operating Costs (COC)

In simple terms, DOC (or Hourly Flying Costs) includes all those costs which are associated with and dependent on the type of aircraft being operated and which would change if the aircraft type were changed. IOC are all those costs which remain unaffected by a change of aircraft type (e. g. station and ground costs, ticketing, sales and promotion costs, general and administration costs). The Cash Operating Costs is part of DOC and effectively consists of the crew and fuel costs aspects of the flying costs.

The elements of aircraft operating costs are:

- Fuel costs
- Cockpit crew costs
- Airframe maintenance costs

- Engine maintenance costs
- Cabin crew costs
- Landing fees
- Navigation charges
- Insurance
- Depreciation
- Interest
- Administration costs
- Marketing costs
- Staff costs

There are numerous methods of estimating direct operating costs, each tailored to the specific needs of an airline. An airline will group the above elements into direct and indirect costs depending on the method selected. Some of the methods are listed below:

- ATA – 67: Air Transport Association (1967)
- AEA: Association of European Airlines (1989)
- Piano97 method (Lissys Ltd)
- AI: Airbus Industrie (Commercial)
- AI: Airbus Industrie (Project)
- Boeing
- MDD: McDonnell Douglas
- RR: Rolls Royce
- Lufthansa: Airline
- FEAT: Future Economic Advanced Transport WG
- SAWE: Society of Weights Engineers
- CEE: Committee Etude Economique

The ATA – 67 method was updated in 1967 to take into account turbine powered aircraft. It is a standardised method for estimating operating costs and provides a means of comparing operating economics of competitive aircraft under a standard set of conditions.

The AEA method is based on long established procedures of the AEA (1989 release), which exists in two variations, 'short-medium range' and 'long range' aircraft types. The AEA currently has a membership of 35 European airlines.

The P97 method is another method for estimating direct operating costs used by a major airframe company unknown to Stirling for project analysis. This exists in four versions, 'short', 'medium', 'long' and 'extra-long' range aircraft types.

Cost Index (CI) is the ratio of time-related direct operating costs to fuel price. This figure, when input into the aircraft flight management system (FMC), determines a number of flight parameters such as rate of climb and airspeed that can affect the fuel economy of the flight.

This paper comprises a study of the Cost Index and how airlines use it strategically and operationally, and what opportunities there might be for product differentiation by aerospace manufacturers by considering CI more explicitly at the design stage. The ultimate aim is to enable aerospace manufacturers to identify options for product differentiation.

With the recent introduction of the European Emission Trading Scheme (EU ETS), aimed at reducing the emission of greenhouse gases, fuel conservation techniques became even more important as there is a direct relation between CO_2 emission and fuel burn; however reducing DOC may still be the most important thing with the addition of buying and selling carbon credits.

However, it has to be noted that despite the growing pressure on airlines to minimise their fuel costs, any implementation of fuel conservation techniques has to comply with the existing safety regulations. Any fuel efficiency programme should never compromise safety.

The remainder of this document is structured as follows:

- Section 2 is essentially a tutorial on airline cost practices and describes the range of methods used by airlines globally and various options of the measurement of individual cost items.
- Cost of fuel, particularly fuel cost index, its usage in practice and various initiatives within the aerospace sector are described in Section 3.
- Section 4 describes the functionality of the systems currently available on the market, namely flight management systems, flight planning systems, and dedicated cost index calculation systems.
- Section 5 describes CI Strategy and the different decision processes and strategies between traditional and low-cost airlines.
- Section 6 describes the manufacturer's considerations during the design phase and how this can affect direct operating costs and cost index. Potential opportunities are briefly identified. This area of the study is the focus for the on-going work and more details will be provided in future studies.
- Section 7 describes some case studies where Stirling has been involved in

work to improve aircraft performance and reduce elements such as fuel consumption, turnaround times and maintenance costs.

- Section 8 describes conclusions and recommendations.

2 Airlines' costing practices

This section describes airlines' current practices regarding the determination of their Total Operating Costs (TOC). TOC represent all costs which can be allocated to an individual aircraft movement①. The most common approach to analysing TOC, in general, is to categorise and assess the various components according to their cost behaviour.

Understanding these costing practices at the macro-level is important in order to identify opportunities for improving fuel cost analysis and control at the micro-level.

2.1　Cost behaviour approach to cost analysis

'Cost behaviour' refers to the manner in which costs vary in response to managerial action (i. e. in this case, decisions taken with respect of output volume②). [3] TOC is normally divided into the following categories:

- Direct Operating Costs: DOC include all those costs which are associated with and dependent on the type of aircraft being operated and which would change if the aircraft type were changed. DOC can be further divided into:
  - ＊ Variable costs: Variable DOC depends also on the output generated by the aircraft (e. g. fuel costs, landing charges, variable components of crew salaries).
  - ＊ Fixed costs: Fixed DOC remains the same regardless of the output generated by the aircraft (e. g. depreciation of aircraft, fixed component of crew salaries).
- Indirect Operating Costs: IOC are all those costs which remain unaffected by a change of aircraft type (e. g. station and ground costs, ticketing, sales and promotion costs, general and administration costs). These are therefore by definition not directly dependent on aircraft operations, neither in terms of the amount of generated output nor in terms of aircraft type used to generate this output. [1]

---

① Airlines can also sustain certain non-operating costs, such as interest expenses, affiliates' losses, or loss on asset disposals. However, these types of costs are out of the scope of this study.

② Output volume is the amount of provided services measurable e. g. by number of flights, or available seat kilometers.

This paper focuses on DOC only as these are the costs which can be affected by the type of aircraft or the design of a future aircraft. A high-level summary of various types of DOC together with their evolution in the last 30 years is shown in Table 1. Cost of fuel has historically accounted for 10%– 15% of TOC. However the current trend of rising fuel prices and streamlining other cost items raises this figure to 20%– 40%. [1][3] The cost of fuel has trebled over the period between 2002 and 2011 (the cost of fuel in 2002 was 0.72 $/gallon, whereas the value in 2011 was 3.01 $/gallon).

Table 1   Classification of DOC and evolution of the percentage of total operating costs[2][1][2]

Direct Operating Costs (DOC)		Percentage of TOC			
		1980	1990	2000	2010
Costs of flight operations	Labour costs of flight crew	8%	7%	8%	9%
	Fuel and oil costs	28%	15%	11%	25%
	En-route and landing fees	5%	4%	7%	7%
	Aircraft insurance and rental	3%	6%	9%	7%
Maintenance and overhaul costs	Engineering staff costs and maintenance material	11%	11%	11%	10%
Depreciation and amortisation	Depreciation of flight equipment, extra depreciation	6%	7%	7%	6%
Total		61%	50%	53%	64%

Whilst Table 1 shows an example of the categorisation of DOC items, there is no universal agreement on how to allocate costs between DOC, COC and IOC. In fact, cost structure can vary considerably from one airline to another depending on the model selected, which is one of the reasons that high level inter-airline comparisons are usually of limited value. A huge factor in the percentage breakdown is the economic climate at the time as this can lead to big changes in fuel and oil for example. Below are a couple of examples of the percentage breakdowns used by other models (see Fig. 1, Fig. 2, Fig. 3 and Fig. 4).

---

① In May 2008, IATA's website quoted from Platts a jet fuel price index of 391.4, based on 100 in the year 2000. [3]

② The figures represent scheduled airlines only.

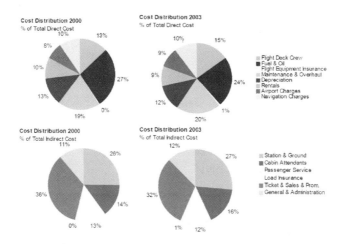

Figure 1   Operating economy of AEA airlines cost distribution[28]

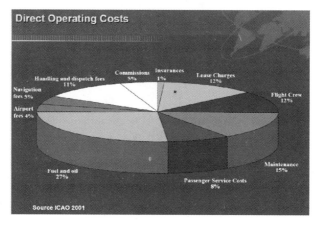

Figure 2   DOC element of cost distribution
ICAO 2001[29]

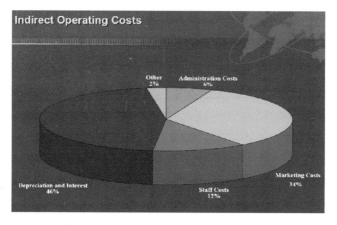

Figure 3   IOC element of cost distribution
ICAO 2001[29]

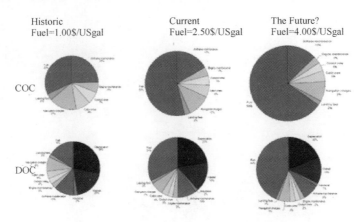

Figure 4　Airbus COC and DOC example charts[30]

Although there is not an industry-wide agreement on the allocation of individual cost items into various types of costs，（ⅰ）fuel cost is universally classified as DOC and（ⅱ）there are four levels of decision-making generally utilised by airlines，which can be used as a basic guidance in terms of the categorisation of costs：

1　Industry entry decisions：A decision to enter the airline industry means，among other things，the establishment of a corporate infrastructure，which generates general and administrative overheads.

2　Network design，scheduling，and other product decisions：These generate fixed fleet，crew，maintenance，and station costs，as well as a marketing and service delivery infrastructure.

3　Flight-related decisions：Any decision to operate a flight，whether in accordance with a schedule or not，leads to expenditures which usually fall into variable DOC. Notable amongst these are fuel，ATS，airport，handling（passenger and ramp），and variable crew and maintenance costs.

4　Passenger-related decisions：Acceptance of each individual passenger onto a flight with available seats may imply costs such as travel agency commission，global distribution systems fees，catering，and marginal fuel-burn. [3]

As a result of this decision-making guidance，most IOC and fixed DOC fall into levels 1 and 2，whilst variable DOC will go into 3 and 4.

The percentage of DOC varies considerably among the airline business models. Airlines with minimal sales，distribution，and promotional expenses（such as charter airlines and low cost carriers）might have as much as 80% of TOC comprised of DOC. DOC at traditional，scheduled air carriers usually account for 50%- 70% of TOC. DOC is also driven by external（largely uncontrollable） factors such as fuel price (of direct interest to this study) and labour cost.

Although DOC may remain the same if the aircraft does not fly, the IOC will still be applicable leading to an increase in TOC. If the airline produces an output greater than zero, but within the range of its existing capacity, there will be an increase in TOC only by the added variable DOC. Therefore it can be assumed that unless efficiency can be improved by changing the operating system in a way that alters the relationship between output and cost, the TOC rises as the output rises.

One of the major objectives of the corporate restructuring undertaken by a number of airlines in recent years has been to alter their current output/TOC relationship by becoming more efficient. Technology has historically played a major role in improving this relationship (e. g. more efficient airframes and engines, online distribution). However, reorganisation of the value chain and of working practices within the airline has made a growing contribution in the last 15 years. [3]

Another huge factor that can affect operating costs and the percentage breakdown is utilisation of an aircraft. Utilisation is defined as aircraft days per block time. The more routes an aircraft flies per day the more the fixed costs can be spread across a higher number of trips. In reality it is not straightforward to just increase utilisation in an effort to reduce flight costs as utilisation is limited by such factors as legislation (time required for the crew to rest before the next leg), airport operating times (Take-off and landing restrictions at certain time of the day) and customer requirements (preference to fly at night). This change in fixed costs has an effect on fuel being the biggest factor contributing to direct operating costs. The airline's decision to purchase or lease an aircraft will also affect DOC.

Short haul utilisation for EasyJet and IAG (British Airway and Iberia) was 11. 3 hrs/day and 8. 5 hrs/day respectively in 2011. These factors are highlighted later in the paper when comparing the CI strategy between traditional and low-cost operators (see section 5).

2.2 Alternative approaches to cost analysis

Other than 'cost behaviour' based analysis of costs, there are several alternative approaches currently utilised by airlines in addition to, or instead, of this approach. These are:
- Cost analysis by department: This approach identifies costs according to the airline's organisational structure and is akin to the traditional cost and profit centre approach.
- Cost analysis by product: This approach is based on the philosophy that

each product should be charged not only with the direct costs that went into producing and delivering it, but also with an allocation of the indirect costs. However, there are two inherent issues with this approach — identifying the product (e. g. by geographical region, by cabin class, or by fare type) and allocating costs.

- Cost analysis by route: This approach is based on the allocation and analysis of all costs connected with a particular route. Airlines may face complications when trying to analyse connecting flights or origin-destination pairs, which are served by other than a non-stop flight. These services have to be analysed by aggregating the TOC from each of the respective routes up-stream and down-stream.

- Cost analysis by fleet and sub-fleet: Most airlines use this type of cost analysis mostly when comparisons between aircraft types are being drawn for fleet planning purposes. It is also important to the fleet assignment process — where different aircraft types could be assigned to a route, the choice should, in principle, be the type that maximises the gap between operating revenue and operating cost given the particular schedule.

- Cost analysis by activity: Activity-based costing collects costs by activity and then allocates shares of each activity's aggregate cost to products, markets, departures, stations, fleets, individual aircraft, cycles, entire schedules, and other cost groups on the basis of the amount of each activity the cost group consumes. The purpose is to assist airline managers in making cost-effective decisions by identifying costs with activities and isolating non-value-adding activities.

- Cost analysis by contribution: Any excess of revenue over variable DOC can be defined as a contribution made by the product, route, department or other profit centre concerned towards coverage of the airline's fixed DOC and IOC. Services are therefore assessed according to their contribution to these costs. [3]

Since the mid-1990s, major airlines have been using increasingly sophisticated models to analyse both the cost implications of different output decisions and areas where operational efficiency could be improved. [1][3]

---

① Various generic cost models are currently available from industry associations and the leading airframe manufacturers. Although these models have proved to be useful, particularly when comparing the operating costs of alternative aircraft types, individual airlines have different cost structures and therefore only airline-tailored models are usually used for profit planning purposes. [3]

Case study

British Airways began looking at activity-based costing in the late 1980s, but relatively few carriers have followed this approach. Many of these airlines do not have management accounting systems aligned with activities. To realign from existing, predominantly departmental, accounting systems would be expensive and also politically sensitive. Furthermore, activity-based costing relies heavily on surveys and estimates, and the accounting systems can be expensive to operate. [4]

2.3 Measurement of costs

The industry best practice is to measure all the operating costs on a unit basis. Unit cost is defined as total operating cost divided by airline's output, and is therefore expressed as cost per available seat kilometer (ASK) or available tonne kilometer (ATK). [1] Depending on the measure (ASK or ATK), unit cost can be defined as cost per available seat kilometer (CASK) or cost per available tonne kilometer (CATK):

$$CASK = \frac{TOC}{ASK} \quad CATK = \frac{TOC}{ATK}$$

CASK is basically the cost of flying one passenger seat on a given route. Similarly, CATK gives an overview of costs per one available tonne of payload. In general, CATK is more widely used because of its application also to passenger flights (usually under the assumption that a passenger plus his/her luggage has a mass of 100 kg) and hence better comparability with other airlines. However, it must be noted that unit costs are rarely the same at different levels of output (e. g. route, station, network levels).

Besides CASK/CATK, airlines also use other key performance indicators (KPIs) to measure their efficiency and level of costs. The most important KPIs connected with an airline's costs are the following:

- Revex Ratio: Revex Ratio (RR[2]) describes how much revenue has been earned by each unit of expenditure. The ratio is commonly expressed in

---

① ASK/ATK are basic measures of airlines' output and performance. ASK is equal to the number of seats available on a particular route multiplied by the number of kilometers flown. Similarly, ATK is equal to the maximum payload available on a particular route multiplied by the number of kilometers flown.

② RR not to be confused with Rolls Royce

percentage：①

$$RR = \frac{OR}{TOC} \cdot 100$$

$RR = 100\%$ represents a break-even point where operating revenues equal operating costs. Figures higher than break-even point show the fact that the airline (or particular route/station/region) generates an operating profit.

- Operating margin：Operating margin (OM) represents one of the most important macro-level KPIs. OM is defined as a rate of operating profit ($OP = OR - TOC$) to operating revenues as a percentage figure：

$$OM = \frac{OR - TOC}{OR} \cdot 100 = \frac{OP}{OR} \cdot 100$$

- Break-Even Load Factor：Break-Even Load Factor (BELF) is the load factor at which total operating costs equal the revenue from the sale of seats and is calculated as a percentage：②

$$BELF = \frac{\dfrac{TOC}{ATK}}{\dfrac{OR}{RTK}} \cdot 100 = \frac{CATK}{Y} \cdot 100$$

Since operating costs vary from one airline to another, so does the break-even load factor. Increasing DOC results in an increase in the BELF. In recent years the industry standard has increased to approximately 70%.

The configurations of aircraft and the utilisation of these aircraft by an airline can have a huge impact on the BELF. An increase in seats available of course will lead to an increase in revenue and hence a lower BELF.

3　Cost of fuel

As shown in Table 1 (Section 2), the most significant item of DOC today is the cost of fuel. The most important fuel cost drivers are：

- The age and fuel efficiency of an airline's fleet；
- The market price for jet fuel and the capability of the airline to hedge；
- Regional market price pressures；
- Network design；

---

① OR＝operating revenues

② Y＝yield (revenue per unit of output sold)；RTK＝revenue tonne kilometers (payload sold on a particular route multiplied by the number of kilometers flown)

- Local factors at airports on an airline's network; and
- Exchange rates. [3]

Although some of these drivers are uncontrollable, airlines can exert a degree of control over their fuel-related expenditures. Given the current high fuel price, even marginal improvements may be significant. The following steps can be taken by an airline to mitigate the impact of fuel price on its performance:

- Hedging: Airlines tend to hedge against the fluctuations of fuel prices by contracting a fuel supplier for a certain period at a particular unit cost. The airline can hedge completely or just partially (e. g. Southwest hedged 70% of its fuel needs in 2008, but United only 16%). However, fully hedged airlines are rare in practice (due to the common practice of spreading their risks). Other than major air carriers, and especially airlines with weaker performance, may have problems with establishment of hedging contracts because of their financial standing. In addition, hedging brings several risks, such as ending up paying more than the market price for fuel currently is (e. g. the case of Delta in 2006).

- Purchasing: Some airlines cooperate when purchasing fuel from a supplier in order to gain higher rebates.

- Fleet modernisation and modification: New aircraft types with fuel-efficient engines, aerodynamic refinements, lighter weight (attributable particularly to greater usage of composites and more efficient systems ultimately reduce the fuel costs). However, exterior and interior modifications can also help to increase the fuel efficiency of an aircraft (e. g. Frontier's investment in lightweight seats for its Airbus fleet in 2007 was expected to save US$5 – 6 million per annum at then-current fuel prices and aircraft utilisation levels).

- Aircraft configuration: An aircraft that has an all economy layout will have lower DOC than a typical 3-class configuration however the revenue generated for each flight will be different (this will affect the BELF).

- Operational practices: The potential internal fuel-saving measures connected with operational practices include:
  * Reducing aircraft weight (e. g. by cleaning, or implementing weight reduction programmes during heavy checks);
  * Eliminating unnecessary passenger service items and ensuring that passenger and hold baggage weights comply with approved standards;
  * Optimising the aircraft's centre-of-gravity and trim in order to minimise cruise drag;

> * Ensuring that flight controls are properly rigged to avoid misalignments;
>
> * Checking the accuracy of instruments (particularly Mach-meters);
>
> * Repairing dents, damaged seals, panel misalignments and other surface irregularities to minimise the airframe drag;
>
> * Regularly washing engines to maintain the aerodynamic efficiency of compressor blades;

- Using economic tankering programmes to provide flight crew with real-time guidance in respect of optimum purchase locations (used mostly in the case of short-haul flights);

- Stipulating and monitoring operating procedures for: reduced engine taxiing (e. g. single-engine taxiing on a twin or two/three-engine taxiing on a quad engine aircraft), take-off (e. g. derated thrust, lower flap setting to minimise drag), initial climb (e. g. cleanup as soon as possible), step-climb and cruise (e. g. using area navigation (RNAV) to fly user-preferred direct routings where onboard equipment permits, and keeping as close as possible to the aircraft's optimum altitude at different stages of the flight given its reducing weight and prevailing weather conditions), approach (e. g. performing Continuous Descent Approaches (CDAs) or use lower flap settings to minimise drag), landing (e. g. limiting use of reverse thrust), and turnaround (e. g. using ground power units and other sources of airport-supplied power);[1]

- Monitoring actual Fuel Over Destination (FOD) levels to minimise systematic carriage of unnecessary reserves that add weight and, on long-haul legs, might displace cargo;

- Monitoring fleet performance to identify individual aircraft recording above-average fuel consumption; and

- Planning flights carefully to optimise horizontal and vertical profiles consistent with day-of-operation weather, optimise the aircraft's fuel cost index, select suitable alternates to minimise carriage of reserve fuel, and where appropriate use en-route redispatch via an intermediate point on long-haul flights to reduce fuel reserves relative to those required for pre-flight dispatch direct to the final destination. [3]

The following sections deal with the fuel cost index — a state-of-the-art approach of optimising the airlines' fuel costs.

---

① These procedures are subject to safety considerations, local and air traffic conditions.

3.1 Fuel cost index

Proper implementation of the fuel CI is one of the most promising techniques to thoroughly understand the cost of fuel and to optimise DOC. The CI used in the Flight Management System's (FMS) managed mode provides a flexible tool to control fuel burn and trip time to get the best overall economics of the flight (Section 4.1 describes FMS functionality in more detail).

However, in general, any technique that reduces fuel burn necessarily means longer flight time. Hence fuel savings are offset by time-related DOC and also may be limited by the availability of slots at the destination airport.

In order to estimate the most appropriate climb/cruise/descend speeds for a particular route or flight bearing in mind both the fuel as well as time-related DOC the CI is defined as the ratio of the time-related DOC (DOC_T) for an aircraft operation and the associated cost of fuel (C_F):

$$CI = \frac{DOC_T[\$/hr]}{C_F[\$/kg]}$$

DOC_T are all those cost items which are related to block time[①] flown by the aircraft except fuel costs. These are for instance:

- Utilisation of flight crew and route salaries/bonuses;
- Oil costs;
- Overhaul costs (those related to block time);
- Marginal depreciation costs; or
- Aircraft/flight equipment lease costs (based on block time).

In theory, entry of the CI into the aircraft's FMS facilitates the calculation of the appropriate trajectory for the given flight conditions. CIs are entered and an appropriate economical (ECON) airspeed is calculated by the FMS. Compared to fixed-Mach flight planning or Long Range Cruise (LRC) speeds, the CI optimisation of planned speeds can provide savings of approximately 2%–3% and in some cases as much as 10% when a flight is restricted to a low altitude or in unusually strong winds.[11]

In addition, CI-based flight plans should be also available for non-FMS equipped aircraft. While many modern larger aircraft are equipped with CI optimisation embedded in the FMS, many regional jet aircraft and other older generation aircraft do not have the necessary technology. However, CI

---

① Block time is defined as the time from off-blocks at airport A to on-blocks time at airport B for any single aircraft movement.

optimisation is available for these aircraft from vendors of CI systems, which operate independently of the FMS. This technology is currently available as a software application on Class 1 and Class 2 Electronic Flight Bag (EFB) systems and even as flip charts or booklet-based systems. [13]

If fuel costs represent the highest priority for the airline (e. g. because these costs are more significant than DOC_T), then the CI value should be low. In extreme cases, entering zero for the CI results in maximum range airspeed, so called Maximum Range Cruise (MRC)①, and minimum fuel consumption at a particular flight. This setting virtually ignores the cost of time (DOC_T).

On the other hand, if the fuel costs are considerably lower than the other components (DOC_T) the CI value used should be high. Again, the extreme case involves entering the highest possible value (999). The FMS would then use a minimum time speed schedule (i. e. minimising DOC_T by flying at maximum flight envelope speeds) and the cost of fuel is ignored.

The figure 5 shows the ECON speed and its association with C_F, DOC_T, as well as other common operational speeds described before (MRC, LRC).

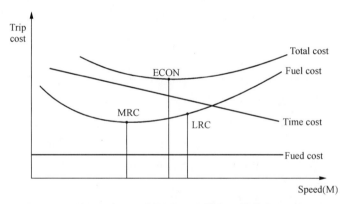

Figure 5　Dependence of ECON, MRC and LRC speeds on fuel costs and time-related costs①

The extreme CI values are in fact not used for real operations. One of the reasons for this is the fact that it is very rare to have such significant differences between C_F and DOC_T (i. e. extremely low fuel costs and high DOC_T, or vice versa).

---

① Also referred to as Mach for Maximum Range (MMR).

② LRC is connected with MRC, which is the speed providing the furthest distance travelled for a given amount of fuel (or the minimum fuel burned for a given cruise distance). LRC is defined as speed above MRC resulting in 1% decrease in distance travelled for a given amount of fuel, which in turn gives 3%–5% higher cruise airspeed. Historically, the majority of long-haul flights have been done on LRC. [5]

Using CI equal to zero is not really advantageous also from the economical perspective as fuel savings are not significant compared to time loss. However, a slightly higher CI gives significant time gains in return for the slightly higher fuel consumption. For example, for the Airbus A319, increasing the CI from 0 to 20 reduces the block time by 15 minutes (5%) for a fuel burn increase of only 200 kg (2%) on a 2,000 nautical mile sector at 35,000 ft (see Figure 6). [6]

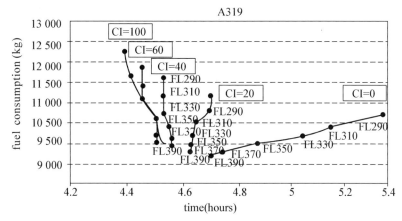

Figure 6   Example of the time/fuel relationship of
particular CI values, Airbus A319[6]

Once the CI is entered into the FMS, the aircraft flies at its ECON speed throughout the whole flight. The current ECON speed at any point of the flight might vary, based on various factors. Figure 7 shows an example of the dependence of ECON speed on flight level for various CIs. The most significant changes in ECON speed with flight levels happen in the case of low CIs, but are almost constant for higher values. Also, the ECON speed is very sensitive to CI changes when flying below the optimum altitude. [6]

Weight is another factor which can affect the ECON speed depending on the entered CI value. Taking an Airbus A310 example (Figure 8), for high CI values, the ECON speed is relatively constant throughout the flight. However, for a low CI, the speed reduces significantly with the reduction of aircraft weight. This is the consequence of the fact that low CI values are more focused on low fuel consumption than on DOC_T. Moreover, a small CI increment for low CI values has a significantly higher influence on the ECON speed, and hence on flight time, comparing to the higher CI values. These trends can be applied to all aircraft. [6]

To summarise, the optimum ECON speed should remain greater than MRC and lower than LRC over the entire range of a typical cruise operation in terms of

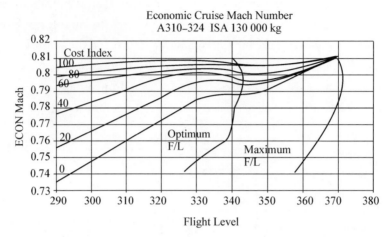

Figure 7　Dependence of ECON speed on flight level
for various CIs，Airbus A310[6]

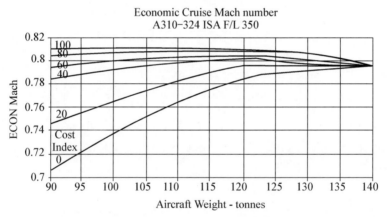

Figure 8　Dependence of ECON speed on aircraft weight
for various CI values，Airbus A310[6]

gross weight and altitude. The optimum flight level for the ECON speed should be the one which provides the greatest specific range at given gross weight. The optimum flight level increases with decreasing gross weights.

In practice, however, the ECON speed is often also limited by various external factors, such as Air Traffic Control (ATC) restrictions or weather conditions. The FMS also limits target speeds appropriately for initial buffet and thrust limits. Furthermore, other factors, such as engine degradation and time between overhauls, limit the use of ECON speeds during take-off and climb (the most fuel efficient take-off and climb procedures are unrestricted full-thrust profiles). [5]

In addition, other potential company initiatives, such as hedging or economic tankering, can make the CI calculations complicated. Therefore the appropriate CI figure varies with each airline and even each flight.

3.2 Airlines' use of CI

In theory, CI optimisation should be the basis for the optimisation of all airline flight operations. CI-optimised profiles burn the least amount of fuel for a given flight time or, conversely, they have the shortest flight time for a given amount of fuel burn. [1] The best way to plan and fly CI-optimised profiles is to use a dedicated optimisation system, both at the flight planning stage and for real-time flight management in the cockpit (currently available off-the-shelf systems are described in Section 4.3).

Since fuel costs, and in some cases, time-related costs vary with the route, the CI should be route specific. Once an airline has made the effort to analyse its DOC_T, then, all other things being equal, the CI for each route should form a basis for schedule construction and day-to-day operations. No other method of flying including fixed Mach, multiple fixed Mach, LRC, or other speeds will result in optimal profile solutions. Furthermore, these basic profiles use only simplistic altitude selection methods. However, altitude is a critical component of the profile. CI solutions solve both the altitude and the air speed issues. [13]

If an airline decides to adopt genuine CI flight management, two possibilities exist:
- Route and aircraft specific airline cost analyses can be performed tailored to the network and its operating and economic environment;
- Aggregate approximations can be performed, bundling routes in low/medium/high fuel and time-related cost brackets, which the airline may decide to adopt as the most pragmatic approach. [13]

In addition to the initial set up, airlines should periodically review the inputs. These should include both C_F and DOC_T. [13]

Using the same CI in pre-flight planning as well as in the FMS is the basic parameter for the airline to provide fuel-efficient operations. In the past, updating the FMS's CI was restricted to maintenance staff, but now some airlines have defined company policy to allow the crew to check and enter the CI independently.

---

[1] It should be remembered, however, that shortest flight time is not always the goal. For example there may be curfews in place at the destination airport, arrivals might be restricted by the slots available to the carrier or arrivals might be coordinated with subsequent departures within the airlines overall network.

The CI value is communicated to the crew via the flight planning document for the specific aircraft concerned. [15]

Although the CI methodology should in theory be the best available method to properly assess airline's direct operating costs in relation to the cost of fuel, the practice brings several issues. A number of direct interviews with airline representatives were conducted and the following drawbacks were discovered:

- There is a common uncertainty whether the DOC_T calculations are performed correctly;
- Airlines are often insufficiently informed about the theory behind CI and its full capabilities;
- Dedicated CI systems are expensive and the whole methodology is resource demanding (both financially and in terms of staffing); and
- Airline pilots are often not well educated to fully understand all the CI aspects.

It was discovered throughout the interviews that there is no real assurance that the DOC_T values are calculated correctly. This issue is mostly connected with the numerous variables, which come to play when calculating DOC_T items. These factors can change significantly and are usually out of the control of the airline (e. g. ATC restrictions). In addition, currently established internal company policies often lower the impact of CI on the possible cost savings, e. g. if an airline has a company policy to calculate flight crew salaries on the basis of scheduled times and the actual times are taking into account just after a certain amount of delay has been created (e. g. for delays longer than 15 minutes), the flight crew costs are therefore not really connected with the actual time of flight and cannot be involved in the calculation.

The company policy dealing directly with the CI can also bias the usage of this method. One major traditional central European airline confirmed that even though they have capabilities to calculate CI values on a dynamic basis and indeed they do so, the company policy firmly set the minimum CI value. In practice, the calculated CI values are often below the minimum boundary, but have to be increased in accordance with the applied policy. An airline representative clarified this approach by the fact that if they followed the calculated values, their aircraft would be too slow and would be disadvantaged against their competitors on the same routes.

Another issue is the level of awareness among airlines. Although the majority of airlines are aware of CI, many are not sufficiently informed to make the best use of it and its potential is often not fully utilised. Comprehensive application of CI is fairly complicated and would need dedicated staff and transparent company procedures, which would enable tracking of all the DOC_T associated with a particular flight. Therefore, even major airlines do not utilise a dynamic calculation of CI, i. e. they stick with one value for one aircraft type over several years irrespective of the fuel price fluctuations or other factors. For example, Ryanair has been using a common CI (of 30) for all flights for the last two years (as of 2nd November 2011).

There are probably only a very few airlines that calculate dynamically CI on a route by route basis. These airlines have usually the following departments involved in the CI calculation:

- Commercial Department: The Commercial Department calculates and provides DOC_T figures.
- Technical Department: The Technical Department inputs values about the current state of each aircraft by its tail number. Airlines can then calculate the most actual parameters of the aircraft and hence the most accurate CI value.
- Flight Operations Department: The Flight Operations Department provides actual fuel prices on chosen airports (the fuel prices are sometimes provided by a dedicated task group). Flight Dispatch calculates the CI itself. Dedicated CI calculation systems are fairly complex and airlines usually enable only officially trained staff to use the software.

The above mentioned input data enable airlines to calculate CI values for each particular aircraft. In addition, some of the world major airlines utilise different CI methodologies for different types of route, i. e. they calculate CI values differently for flights to major intra hubs in comparison with inter hubs, primary airports, or regional airports. The rationale behind such differentiation is the fact that the cost of delay is different for airports with different structure of flights. [1]

It can be assumed that in general small to medium (and some large) airlines do not use the CI to their best advantage or appropriately. The situation is particularly complicated for charter airlines flying ad-hoc routes. This would include conventional charter but also luxury air taxi services such as those operated by Net Jets and Signature. These airlines do not have sufficient staffing resources

---

[1] The cost of delay is larger for hub & spoke services at hub airports comparing to peer to peer flights to regional airports.

nor do they have a possibility to calculate CI dynamically due to the time pressure for finalising and sending out a quote to a potential customer. In fact, the calculation of fuel costs for pricing purposes is often done in a very simplistic way, as shown in the case study below.

Case study

Personnel of the Commercial Department of small to medium charter airlines usually work with fuel prices provided by the Flight Operations Department (e. g. dispatching). These prices are often collected on a monthly basis and represent public commercial prices for the kilogramme of fuel at a particular airport (small and medium charter airlines do not usually have special rates or hedging agreements with fuel suppliers). Employees then simply calculate whether it is more economical to take more fuel in one of the destinations and carry extra load/burn more fuel during the subsequent flight, or they split the refuelling to half (i. e. 50% fuelled at origin and 50% at destination airport).

Despite the various initiatives and services offered by aircraft manufacturers as well as other international organisations (see Sections 3. 3 and 3. 4), commercial pilots are still not always comfortable with using CIs and, indeed, they sometimes even do not have a proper knowledge about their usage and set up.

For instance, they often do not have a clear idea of what is the difference between the ECON speed and other types of speed, such as LRC. One of the common misunderstandings occurs when a pilot faces a low fuel situation at destination. In this situation they tend to switch to LRC thinking that this speed gives them the longest possible distance flown given the rest of the available fuel. However, the best fuel conserving strategy is to select a very low CI (in fact, even $CI = 0$ is possible). In addition, LRC is usually not adjusted for current/forecast winds at various altitudes compared to ECON speed. [1] LRC is therefore in theory ideal just for zero wind conditions, which in practice happen very rarely. [5]

Aircraft captains also tend to increase the CI in order to increase speed and mitigate delays, especially in the current culture where punctuality is (i) a service necessity in terms of, for example, connecting flights; and (ii) a service differentiator. While some specific situations are covered by company policy, such

---

[1] FMS reduces ECON speed in the presence of tailwind and increases the speed during headwind.

increase of speed is mostly counterproductive. Pilots burn a considerable amount of fuel but reduce a delay by just few minutes.

Contrary to the common belief of many pilots, the variation of ECON speed with gross weight variation (due to fuel burn) at a given flight level is very small. It is hence always possible to fly at a fixed CI which results in a negligible speed variation. [15]

It is therefore clear from the examples mentioned above, that the awareness of CI benefits and usage among airlines need to be enhanced.

3.3 Aircraft manufacturer initiatives

Aircraft manufacturers have been and continue to strive to educate airlines and their pilots to maximise the use of CI. In particular, Airbus and Boeing have been very active in this field recently: Boeing via its 'Aero' magazine and the offer of consultations, Airbus through its 'Getting to Grips' series of educational papers.

Fokker is also trying to educate its potential customers with reports on this topic, particularly Fuel and Environmental Management report released as a part of a Fokker Services service portfolio. [14]

3.4 Initiatives of other organisations

In addition to aircraft manufacturers, the International Air Transport Association (IATA) is also active in the education of its members. IATA has been supporting airlines by launching industry-wide initiatives under the Fuel Action Campaign. Fuel Action Campaign consists of four components:
- Save One Minute initiative dealing with introducing better airspace design;
- Route Optimisation initiative to operate more direct routes;
- Improved Air Traffic Flows and rationalisation of existing noise abatement departure procedures; and
- Fuel Conservation through efficient operating procedures.

Fuel Conservation initiative aims to introduce a 1% improvement in fuel efficiency across the industry. IATA is compiling industry best practices, publishing guidance material and establishing training programmes for its member airlines to improve existing fuel conservation measures. [10]

IATA also offers Fuel Efficiency Gap Assessment (FEGA) for IATA member airlines and Fuel Efficiency Consulting (FEC) for non-members. Both are optional projects followed by implementation projects supported by IATA experts. There are currently five different teams, each composed of three experts: a flight operations expert, a maintenance & engineering expert and a flight dispatcher. The dedicated teams calibrate fuel efficiency calculator, develop CI and debrief airlines on on-site visits with first applicable suggestions. In addition, the off-site analysis of collected data and delivering of a final report is also a part of the service package. [12]

As of May 2007 (which is the latest data available), over 50 airlines had been visited by IATA teams since September 2005. Total savings achieved have been reported at more than a US\$1 billion, representing in average 2 – 12% of total fuel budgets. 43% of the total potential savings arise in flight operations, 37% in the dispatch area and 20% in maintenance and engineering. [12]

4   Functionality of the systems currently on the market

4.1   Flight Management Systems

Most modern aircraft are equipped with Flight Management Systems (FMS) of different levels of sophistication. Although the basic FMS functionality has not changed since the advent of glass cockpits, some of the high-end FMSs provide the CI optimisation capabilities as well as extremely accurate time and fuel predictions. [13] These systems are able to optimise flight plans for winds, operating costs and suggest the most economical cruise altitude and airspeed, depending on the CI chosen by the airline.

Other FMS have only basic capabilities with no speed or CI optimisation. Some older FMS versions will recommend a flight level based on weight regardless of winds or CI values. In the case of older generation or regional aircraft without FMS altitude information available, the Aircraft Operating Manuals simply recommend altitudes normally based on weight for LRC speeds (no wind or CI input). [13]

Irrespective of the type of the FMS used, the accuracy of the information entered in the FMS determines the performance of the system's information and predictions, and hence the efficiency of the whole flight.

The FMS can operate in two modes – managed and autonomous. When in the managed mode, the flight crew interfaces through the Multipurpose Control and Display Unit (MCDU) and enters basic flight variables such as weight, temperature, altitude, winds, and the CI (if available). From these data, the FMS computes the various flight control parameters such as the climb law, step-climbs, ECON speed, optimum altitude, or descent law. Hence, when activated, this mode enables almost automatic flight management – aircraft performance data are extracted from the FMS database and applied directly on the performance factors. The FMS database is simplified to alleviate computation density and calculation operations in the FMS, but individual aircraft performance factors can produce good correlation with actual aircraft fuel burns. [15]

When in the autonomous mode, the FMS serves as an ideal means of performing cross-check of the flight plan time and fuel data at the pre-flight programming level. While many pilots have different methods of performing fuel checks during flight planning, many limitations exist. Fuel performance charts will only consider data provided by manufacturers, and therefore items, such as aircraft specific airframe and engine in-service deterioration, CI, winds and temperatures at specific waypoints, or last minute Zero Fuel Weight changes cannot be considered. [13] The issue is that these are only approximate methods and there is no precise method for cross-checking the accuracy of the flight plan.

However, inserting the most accurate available information is crucial. For instance, during the pre-flight checks pilots should insert data about the departure runway, Standard Instrument Departure (SID) with appropriate transition, the planned route with the planned arrival procedure (e. g. Standard Terminal Arrival Route (STAR)), and the planned runway. For those FMS, which are capable of calculating the CI, it is critical to enter the winds and temperatures at each waypoint (ideally downloaded directly from the flight planning system, see Section 4. 2). The altitude step-climbs (or descents) should be also entered as these will be used by the FMS to compute additional wind predictions and times. [13] If the FMS is used also for the prediction of optimum altitude, the winds above and below the planned cruise flight level have to be inserted as these will be taken into account by the FMS when determining the fuel efficiency of potential altitude change.

Once all of the available information has been inserted, the minimum FOD, which should include the regulatory final holding fuel (30 – 45 minutes) as well as the alternate fuel, should be subtracted from the planned FOD to determine the amount of discretionary fuel for the trip. In addition, the in-service performance deterioration factor (drag factor) of a specific aircraft should be also entered to the FMS to provide increased accuracy. Improperly programming the FMS may lead to crews wanting to add fuel to compensate for inaccuracies. [13] This can be costly especially on long-haul flights where it can impact not only the overall weight of the aircraft and hence its fuel burn, but also the maximum value of available payload.

Once the aircraft is airborne, the FMS manages cruise speed according to the aircraft gross weight, flight level, wind and entered CI. The flight crew should continuously monitor the FOD and estimated time of arrival (ETA) and cross-check them with the given flight plan. Any differences should be reconciled. If a high CI value was planned for the flight and it turned out that the FOD falls under the minimum level, the CI should be reduced to ensure that adequate fuel is available on arrival. If the flight is held at a less than optimal altitude, FMS should be allowed to compute the ECON speed for the current altitude to minimise the fuel burn. [13]

On long-haul flights, more recent wind and temperature data should be entered to the FMS after a certain airborne period. Also, when the final cruise altitude is different from the planned altitude, wind and temperature information for the new altitudes should be inserted. [13]

Therefore, the flight crew should not change the CI value in the case of different en-route winds, but just update the wind data. However, the CI should be changed in flight if the following situations occur:

- The encountered winds are becoming so great that it can result in a missed hub connection upon arrival. The adjustment of CI value should be done only after checking the resulting FOD value.
- Flight crew can also change the CI in the case of fuel problems (e. g. to avoid performing a refuelling stop). This can happen because of the strong winds encountered or ATC rerouting, restrictions or expected holding at arrival.
- Changing the CI for speed control should never be done except in the case of fuel problems (LRC or MRC), as explained above. [15]

Case study

Fokker 70 and Fokker 100 aircraft are equipped with a FMS capable of efficient pre-flight and en-route flight planning. The FMS enables the crew to recalculate in-flight the optimum vertical profile using the three different options:

Minimum Time;

Minimum Fuel; and

Economy (balanced fuel and time related cost using the CI).

After activation of the Economy performance function, the cost optimisation process is started. The ECON speed is obtained from a table stored in the FMS. The ECON speed is pre-calculated during an off-line study. This study is based on the minimisation of the cruise cost function (λ), representing the cost per unit of cruise distance:[1]

---

[1] FF_{CRUISE} = the total cruise fuel flow; V_G = the airplane ground speed

$$\lambda = \frac{(CI + FF_{\text{CRUISE}})}{V_{\text{G}}}$$

Minimisation of λ with altitude and airspeed as the independent variables, results in the minimum cruise cost condition at given values of aircraft weight, wind speed and CI.

After selection of the optimum cruise Mach at a given altitude by the FMS, the corresponding cruise cost is calculated. This process is repeated for different altitudes to determine the optimum cruise altitude. Once the cruise optimisation has been carried out, the climb cost and descent cost are minimised by selecting the pre-calculated ECON climb and descent speeds stored in the FMS database.

When the Minimum Fuel strategic mode is selected by the flight crew, the FMS use the CI value equal to zero. A high value of the CI of around 100 kg/min will be used by the FMS if the Minimum Time strategic mode is selected.

For turboprop airliners, such as the Fokker 50, the optional FMS does not provide en-route vertical profile calculations and fuel predictions. However, guidelines can be prepared in the form of tables or graphs for in-flight adjustments by the crew. Also, EFBs can be used to run software that determines the optimum profile and speeds. [14]

4.2 Flight planning systems

The problem with many flight-planning systems is that the route analysis is based on a fixed Mach number analysis of minimum time tracks. However, this is a very simplistic approach. Planning a flight using the CI and looking at the optimal route laterally as well as vertically to determine the most efficient vertical profile offers great potential savings. An accurate flight planning system should produce the best vertical profile based on the wind data available for each waypoint, the aircraft weight, temperatures and the flight specific CI. [13]

Some flight planning systems offer the possibility to upload updated wind and temperature data automatically into the FMS. [13] Less sophisticated solutions offer manual input of these data into the FMS by flight crew. However, unless the wind information (including winds above and below planned altitudes) and temperatures at the planned waypoints are accurate, the recommended FMS optimum altitude will be incorrect.

Performance advisory systems are available for non-FMS aircraft, which enable the use of CI speeds and altitude optimisation. These systems are available either in a booklet format, electronically as part of the EFB system, or in a stand-alone system. Ideally, the outputs from these systems should be integrated to the flight planning system for greater flight planning accuracy and optimisation. [13]

4.3　Dedicated systems

The following dedicated systems have been identified as the current leaders on the market:

- AirSavings;
- Pacelab CI OPS;
- SITA Flight Planning service;
- Lido/Flight;
- Pilots Performance Advisory System.

4.3.1　AirSavings

Airbus has developed a dedicated software solution in cooperation with f:wz (fully owned enterprise by Sabre Airline Solutions), which deals with economy of flight and the CI in particular. AirSavings is a CI administrator capable of calculating the dynamic custom-tailored CI for every flight together with tankering recommendations.

The database is fed by five data streams from automated interfaces (flight schedule, crew schedule, fuel prices, currency rates, and aircraft rotation) and one interface for manual inputs. Based on these data the appropriate CI per flight is calculated and forwarded to flight planning department of an airline. An example screenshot of the system, particularly city pairs list, is shown in Figure 9.

AirSavings can be obtained in two basic configurations:

- Minimum configuration: Minimum configuration is suitable for small short-haul carriers. It involves web-based browser client. There are no automatic inputs, the system works with single currency only (i. e. no exchange rate updates) and ignores crew costs.
- Full-sized configuration: Full configuration is built on server components (either on dedicated or shared server) and provides all input data via automatic interfaces as well as an output interface to flight planning system. [8]

In addition, AirSavings offer a time dependent maintenance cost wizard (see Figure 10) and a cost-based delay avoidance strategy that manages the cost of delays (e. g. passenger misconnections) and integrates these with the DOC_T to

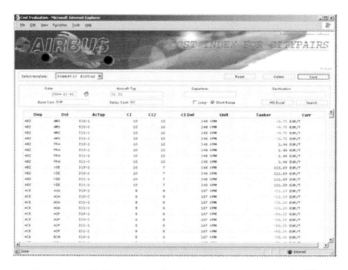

Figure 9   AirSavings application layout, city pairs list[8]

provide CI calculations that can be used to offset late arrival costs. An accurate fuel tankering index for every flight is also provided (assistance with multiple leg tanker decisions is provided as well). [9]

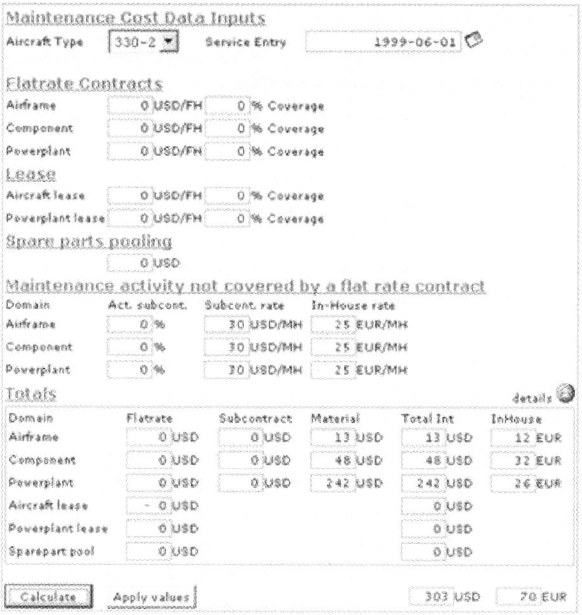

Figure 10   Time dependent maintenance cost wizard, AirSavings[8]

4.3.2　Pacelab CI OPS

Pacelab CI OPS supports the optimisation of vertical flight profiles taking into account airline-specific CI, planned block time and other key factors such as centre-of-gravity and wind. The software supports both the strategic flight planning as well as tactical decisions (see Figure 11).

Figure 11　Pacelab CI OPS application layout[23]

All the data can be imported automatically to minimise the need for manual input. If deviations from the operational flight plan occur, pilots can recalculate the most cost-efficient altitude and speed for the remainder of the trip and adjust the trajectory accordingly.

Pacelab CI OPS runs on commercial off-the-shelf Class 1 and 2 EFBs. The system provides an EFB-optimised touch screen interface, which supports landscape and portrait mode. Intelligent import options allow reading data from electronic operational flight plans (both in proprietary LIDO/Flight format and ARINC 633 standard) and ARINC bus 429 if supported by EFB model.[16]

The system's predefined operating modes cover frequently encountered situations, in which a recalculation of the trajectory is required. These are:

- Preparation mode: The Preparation mode defines the initial trajectory without time constraint based on parameter settings in pre-flight preparation.
- T/O 60s: T/O 60s determines the ECON trajectory based on the actual take-off time and scheduled in-block time (SIBT).
- Flight Level: The Flight Level mode is used for the calculation of the ECON trajectory for the remainder of the trip following ATC clearance of an unplanned flight level.
- Speed: The Speed mode calculates the ECON trajectory for the remainder of the trip following ATC command to change the airspeed at the current flight

level. The mode also evaluates alternative trajectories above and below the current flight level.

- Early/Late: The Early/Late mode optimises the remaining trajectory whenever time targets defined in the original flight plan are not met (with step-climbs and descents taken into account).
- Holding: The Holding mode calculates the most fuel-efficient holding speed for the current altitude and gross weight. [16]

An off-the-shelf version of Pacelab CI OPS is currently available for the following aircraft types:

- BAE AVRO RJ 70/85/100;
- CRJ 200;
- CRJ 700;
- CRJ 900;
- CRJ 9NG;
- EMB 190;
- EMB 195; and
- Boeing 737NG. [16]

4.3.3 SITA Flight Planning

SITA Flight Planning service provides a calculation of fuel for any origin-destination pair. It is the core component of SITA's integrated Flight Briefing portfolio, which also includes the following components:

- Surface Weather and NOTAMs Module; and
- Airport, Runway and Obstacle Data. [24]

SITA Flight Planning service selects the minimum-cost route based on optimum CI and en-route fee analysis. In addition, the system dynamically assesses key factors to determine the true cost of carrying excess tanker fuel. Together with a full profit and loss summary, the Flight Planning system can output this information on the flight plan. A tankering calculation can automatically be considered for every flight plan individually. [24]

Users can create, manage and maintain individually tailored route databases according to their specific requirements. The service can be delivered either via desk-based graphic user interface (using SITA GraFlite software), or via a dedicated SITA Flight Briefing website.

4.3.4 Lido/Flight

Lido/Flight system has been developed by Lufthansa Systems. The company claims that up to 5% fuel savings can be achieved with their solutions. Lido/Flight works with the CI and enables end users to optimise routes either with regard to C_F or DOC_T. Hence, the system is able to identify the best route while taking current flight-related data into account. In addition, Lido/Flight also allows strategic evaluation of future routings and statistical analysis of city pairs.[25]

The flight planning process can be either manual, fully automatic or at any other level of manual intervention. Various flight phases can be analysed and optimised separately and the system provides tailored briefing packages for each flight. Lufthansa Systems provides aeronautical data with world-wide coverage.

In addition to the basic model, several additional modules can be purchased separately. These are:

- Airline Operations Support (AOS): The AOS module supports dispatchers by automating the flight planning process. Airlines can either partially or fully automate their processes for calculating and optimising flight plans and distributing briefing packets to pilots.
- Briefing: Electronic briefing packages enable pilots to access operational and ATC flight plans, fuel summaries, weather charts, satellite pictures, NOTAMs, and other information via the internet. The electronic briefing information can be adjusted according to the airline's needs.
- InflightMonitor: The module continually checks for notifications relevant to all flights which have been calculated but not yet completed and reports them to the dispatcher.
- FreeFlight: This module is primarily applicable in free flight airspaces. It plans the most efficient trajectory for each flight based on the current weather and airspace situation.
- Take-off: The module allows the dispatcher to calculate the maximum allowed take-off weight, considering airport obstacle data, aircraft and weather data. Just before take-off the pilot can also send a take-off request including the final weight, aircraft configuration, and meteorological conditions via data link to Lido/Flight in order to receive a set of optimised take-off parameters.
- Landing: The Landing module combines airport and aircraft-relevant data for an integrated landing performance calculation. It allows the dispatcher to calculate the maximum allowed landing weight for the destination and the alternates for the chosen aircraft and aircraft configuration.
- Traffic Flow Restrictions (TFR): TFR module considers partly restricted airways and calculates the most effective combination of airways between

origin and destination given the availability of these airways. [26]

4.3.5　Pilots Performance Advisory System

Pilots Performance Advisory System (PPAS) is a vertical flight path optimisation tool developed by AASI. The tool is offered in the form of a booklet. It delivers optimised climb, cruise, and descent performance data to the pilot. The data is computed based on CI in the same way as an FMS type system. Performance data are buffet protected and all generated profiles are safe to fly with minimum safe operating speeds for all conditions. The data and presentation are tailored for each aircraft type and operator. [27]

PPAS EFB is an electronic version of the PPAS booklet system. PPAS EFB is available as a Class 1 or 2 EFB systems. PPAS EFB provides a full functionality of the CI optimisation of FMS. It enables the user to dynamically update the system as flight conditions change. The system has four display modes:

- Cruise page: Cruise page shows the optimal solution (ECON speed and flight level), together with supporting information, for a given set of flight conditions, aircraft state and CI (see an example screenshot in Figure 12). The system supports both Reduced Vertical Separation Minima (RVSM) and non-RVSM airspace.

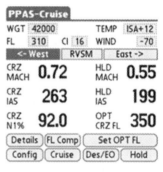

Figure 12　PPAS EFB Cruise page[27]

- Flight Level Compare page: This page enables the flight crew to perform a wind/ altitude trade on a cost basis.
- Hold page: Hold page provides optimal holding speed and flight level data, in addition to buffet margin data.
- Descent page: Descent page provides distance to 10, 000 ft. and the corresponding flight path angle for entry into an FMS, for a chosen descent speed. Engine-out descent speed is also provided. [27]

Flight planning software varies in price and is dependent on:

- Aircraft type
- Fleet size
- Airframe/engine combination
- Airport destinations

Proposals from the software suppliers are detailed in Appendix D.

4.4　Flight Management System and Cost Index Interactions

This section provides a top-level overview of the interaction between the Flight Management System (FMS) and Cost Index (CI). The intent was to provide a lower level case-study view based upon specific systems within applicable short-haul aircraft; however the sensitivity surrounding FMS software has prevented such detailed study (this risk was identified during the initial task scoping).

As described earlier in this paper the CI is the ratio of the time-related cost of an airplane operation and the cost of fuel:

$$CI = \frac{C_{\text{TIME}}}{C_{\text{FUEL}}}$$

The CI is calculated by the airline based on the flight-related costs and cost of fuel. Flight-related costs are:

- Hourly maintenance costs
- Crew and cabin costs
- Marginal depreciation or leasing costs

Hourly maintenance costs are per flight hour. Maintenance costs can be reduced by reducing the flight time, however this means the aircraft has to fly faster resulting in possible increase in wear of the engine resulting in the impact of engine overhauls costs.

Crew and cabin costs are per flight hour. Flight time has an influence on crew cost as this effects rest times and can result in better and more efficient use of crews.

Marginal depreciation or leasing costs are costs such as cost of ownership of the aircraft or aircraft rental. It is based on each extra flying hour rather than fixed calendar time cost.

Fuel costs are the cost of fuel for the trip.

The FMS uses the CI to calculate the ECON Mach speed (minimum cost speed) for the flight profile and is illustrated in Figure 13:

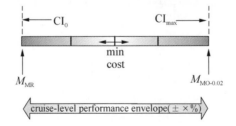

Figure 13　Cruise-level performance envelope

- $CI_0 = M_{MR}$ or MRC (minimum fuel consumption)
- $CI_{MAX} = M_{MO} - 0.02$ (min time and a maximum speed)

In the process of obtaining information on how the FMS uses CI to optimise flight profiles, Stirling made contact with four FMS equipment suppliers, an air traffic/airport management & technology consultancy, pilots from two different European operators and Stirling consultants. As expected the approach to the equipment suppliers to obtain very specific information on their systems was unsuccessful due to data sensitivity. In order to obtain specific information would require a direct relationship between an airframer and equipment supplier.

In lieu of this detail, the following information provides a guide as to how the flight crew enter the flight plans into the FMS (when in the managed mode) to optimise speed as detailed in Figure 14 to create a flight profile:

- A lateral flight plan that defines the intended horizontal flight path
- A vertical flight plan that defines the intended speed and altitude profile for the aircraft to follow while flying the lateral flight plan

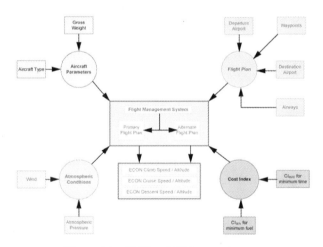

Figure 14   FMS and CI interactions

The lateral flight plan includes the following and is input manually or uploaded from navigation databases (flight planning software can do this, see Figure 15):

- Take-off runway
- Standard Instrument Departure (SID) and transition to en-route mode
- En-route waypoints and airways
- Transition from en-route mode and standard terminal arrival route
- Landing runway with selected approach and approach via

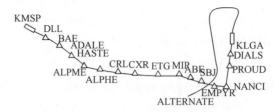

Figure 15   Lateral flight plan example

- Missed approach
- Alternate flight plan

The vertical flight plan gives the FMS all the data required to calculate performance and predictions. Data to be calculated and manually entered by the flight crew into the FMS:

- Cost Index
- Cruise Flight Level
- Zero-Fuel weight (ZFW)
- Zero-Fuel weight center of gravity (ZFWCG)
- Block fuel

The FMS computes the gross weight and the center of gravity (CG) from the ZFW and the ZFWCG entered into the FMS. The gross weight and CG values are used for flight control laws and computation of characteristic speeds. The CG is updated by the FMS as the weight changes during the flight phases.

Weather data to be entered by flight crew or uploaded from databases (measured data):

- Winds (for climb, cruise, descent, approach)
- Seal level atmospheric pressure at destination
- Surface temperature at destination
- Temperature in cruise phase
- The Tropopause altitude

Tactical data for the flight phases (computed by the FMS):

- Phase switching conditions
- Setting of the thrust levers to take-off-go-around
- Reaching acceleration altitude
- Top of climb
- Top of descent
- Passing a deceleration pseudo waypoint
- Touchdown

- Speed profile
- V2
- Economy climb speed or Mach
- Preselected speed or Mach
- Economy cruise Mach
- Economy descent Mach or speed
- Approach speed
- Vertical limitations
- Speed limits
- Speed and altitude constraints, time constraints

Figure 16 illustrates how the FMS uses all these parameters to compute the flight profile using lookup tables.

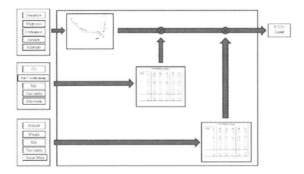

Figure 16   FMS computations to produce ECON speed

This is a general figure to represent how a number of lookup tables are used by the FMS using various inputs. However it must be noted that the tables can consist of different inputs depending on what is being computed and the phase of flight. Variations also exist between different airframe and FMS manufacturers.

Figure 17 is a graphical representation of a lookup table to calculate the optimum ECON speed using weight, flight level, cost index, no head/tail wind and standard ISA conditions.

The lookup tables are generated based on performance calculations that the FMS uses to create the flight profile for a selected mission. To ensure DOC is minimized there needs to be some speed and altitude optimization. This is achieved with specific range as the main driver.

As DOC is calculated per nautical mile, it is possible to plot fuel-related costs, flight-time related costs, and direct operating costs based on Mach number (seeFigure 18).

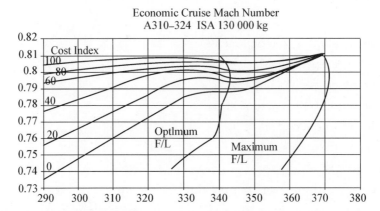

Figure 17　Optimum Mach speed variations based on flight level for each CI

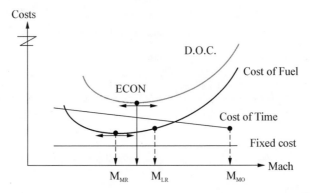

Figure 18　Aircraft performance optimisation in terms of DOC[33]

A flight profile consists of three phases:
- Climb
- Cruise
- Descent

The next few paragraphs discuss how the performance calculations are used to generate the different phases of the flight profile using all the input parameters. It is this performance data that the FMS uses to compute the optimum flight profile depending on the flight plan that is entered into the system.

4.4.1　Climb

The climb phase is the phase of flight that an aircraft climbs to a predetermined altitude after take-off. It is common practice to use a step climb where the aircraft climbs in altitude as fuel burns off.

CI_{MIN} produces a low speed and a high climb angle. CI_{MAX} produces a high speed and a low climb angle see Figure 19.

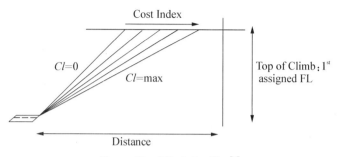

Figure 19   Climb Profiles[6]

An increase in weight, pressure altitude or temperature will decrease the climb gradient and the rate of climb.

An increase in the headwind will increase the flight path angle and decrease the ground distance from top of climb. The rate of climb and fuel and time from top of climb will remain unchanged. An increase in the tailwind will decrease the flight path angle and increase the ground distance from top of climb. The rate of climb and fuel and time from top of climb will remain unchanged. This is illustrated in Figure 20.

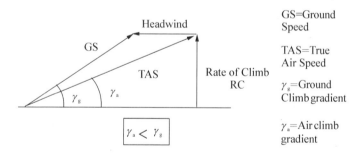

Figure 20   Wind influence on rate of climb[33]

4.4.2 Cruise Flight

Cruise flight is the phase of flight that occurs inbetween climb and descent and accounts for the largest percentage of the trip.

Maximum Range Cruise (MRC) is the cruise speed used when fuel conservation and range is important. MRC can be determined by entering a Cost Index of 0 into the FMS.

Long Range Cruise (LRC) is the recommended speed to minimise trip fuel. LRC is a speed that will provide 99% of the maximum fuel mileage but is slightly faster than MRC. LRC does not take wind into consideration. LRC depends on the actual gross weight and the flight level.

Economy Cruise (ECON) is the cruise speed flown when a valid Cost Index value is entered into the FMS and adjusted for wind.

If the pressure altitude was constant then the weight would decrease resulting in a decrease in M_{MR}, LRC and M_{ECON}. Conversely if the weight remained constant the pressure altitude would increase resulting in an increase in M_{MR}, LRC and M_{ECON}. This is illustrated in Figure 21.

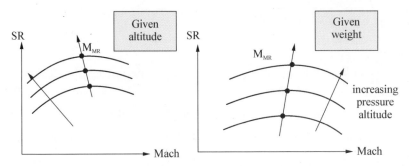

Figure 21　Specific Range as a function of Mach number[33]

An increase in the CI will increase the M_{ECON}, whereas a decrease in the CI will result in a decrease in the M_{ECON}. SR changes with altitude at a constant Mach number, therefore for each weight there is an altitude where SR is at a maximum and is referred as the optimum altitude. At a given pressure altitude, as the weight decreases the optimum altitude and specific range increases. This is illustrated in Figure 22.

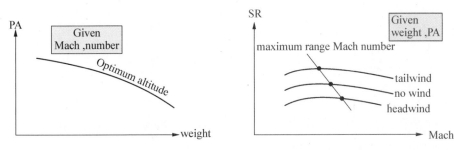

Figure 22　Altitude optimization[33]　　　Figure 23　Wind influence on Mach number[33]

A tailwind will cause the ground SR to increase and the M_{MR} to decrease. A headwind will cause the ground SR to decrease and the M_{MR} to increase. This is illustrated in Figure 23.

4.4.3　Descent

Descent is the phase of flight where an aircraft decreases altitude in its approach to landing.

CI_{MIN} produces a low speed and a low descent angle. CI_{MAX} produces a high speed and a high descent angle. An increase in weight will decrease the descent gradient and the rate of descent see Figure 24.

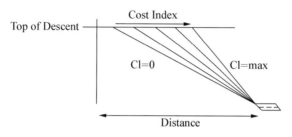

Figure 24　Descent Profiles

An increase in the headwind will increase the flight path angle and decrease the ground distance from top of descent. The rate of descent and fuel and time from top of descent will remain unchanged. An increase in the tailwind will decrease the flight path angle and increase the ground distance from top of descent. The rate of descent and fuel and time from top of descent will remain unchanged. This is illustrated in Figure 25.

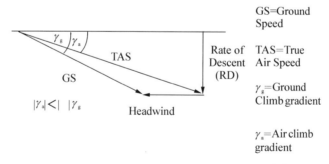

Figure 25　Wind influence on rate of descent[33]

The speed and altitude during the climb and descent phases are limited to the aircraft performance limits and regulatory authorities. To comply with Air Traffic Control requirements, the airspeed during descent is the most restricted speed of all flight phases.

5　CI Strategy

The CI strategy differs from one airline to another. In particular, the difference is greater when comparing a traditional airline (British Airways or

Lufthansa) to a low-cost airline such as easyJet or Ryanair. Although different strategies are adopted, the decisions that result in a certain CI being selected are based on the same elements. The elements to be considered are:

- Elements used in calculating DOC_T
- Fuel pricing and operator strategy
- Schedule requirements
- Route differentiating
- Utilisation
- Aircraft and Engine combinations
- Flight Management Systems selected
- Flight Planning Systems selected

Based on these elements an operator can create a decision tree to determine CI and other performance data for given flights.

5.1 Traditional Airlines

Traditional airlines such as British Airways and KLM-Air France adopt a full-service strategy by focusing on the following elements:

- Selecting centrally located airports
- Extensive network
- Codesharing and mergers
- Mix of aircraft models

Traditional airlines have an extensive global network reached with the use of a mix of single aisle and wide-body aircraft. Traditional airlines operate out of hubs and fly into centrally located airports.

There has also been a trend over recent years where traditional airlines have been merging as well as creating alliances. The advantage of such partnerships for example as the one between KLM, Air France, Delta and Alitalia enables the airlines to share revenues, costs, risk and provides a more extensive, cheaper and efficient network.

Traditional airlines offer connecting flights as part of the airlines existing network or through a partnership with another airline. This is illustrated in Figure 26.

5.2 Low-cost Airlines

Low-cost airlines such as Ryanair and easy Jet adopt a low-price strategy by focusing on the following elements:

- Selecting secondary airports
- Rapid turnaround
- Point-to-point routing

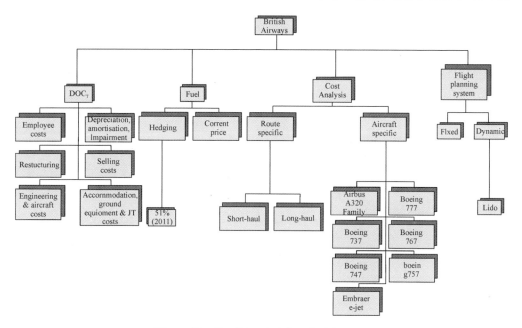

Figure 26   Traditional airline decision tree

- Aircraft selection
- Service
- Staff and overheads

Unlike traditional airlines, the use of secondary airports normally located some distance from major city centres enables the airline to negotiate lower airport fees. Smaller and quieter secondary airports also enable the low-cost airlines to have much faster aircraft turnarounds compared to the industry average increasing aircraft utilisation. Low-cost airline decision tree is illustrated in Figure 27.

The advantage of point-to-point routing means that airlines are able to avoid costs associated with passenger and baggage transfer.

By choosing a single aircraft family an airline is able to simplify processes such as maintenance, reduce the number of operating manuals and hence reduce operating costs further.

Low-cost airlines offer a 'no frills' service meaning passengers had to pay for such amenities as drinks, snacks, checked baggage.

Staff costs are lower compared to a traditional airline with staff generally on a lower salary and having fewer benefits.

Both these examples of decision trees illustrate the decisions a low-cost or traditional airline make when adopting a CI strategy. Fundamentally the same decision is reached when selecting flight planning systems. However the difference

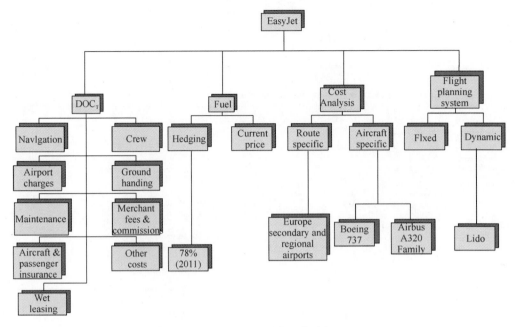

Figure 27　Low-cost airline decision tree

between traditional and low-cost airlines is the different cost strategies that are adopted.

In essence both airlines could operate the same type of aircraft on a given route and even adopt the same CI number. The two decision tree examples illustrate that the elements that make up total operating costs can differ. Fuel strategy is another difference with not all airlines hedging fuel, especially as the fuel price is so high it is common to see airlines adopting a lower risk strategy in not hedging all the fuel that is purchased.

Once the DOC_T and fuel strategy is decided an airline is then able to decide on a CI strategy. Low-cost airlines tend to have one fixed CI number for all flights. This is because they tend to operate a single type of aircraft on short-haul flights into smaller regional airports. A traditional airline like British Airways has a more complex process when selecting a CI. The aircraft used and the selected route influences the CI number. There could also be different CI numbers selected for the different phases of flight.

All of these decisions are made in the flight planning stage and are specific for each airline. A way of reducing the operating costs is with the design of more efficient aircraft and engines. The next section of this paper discusses how considering CI during the design stage can contribute to lower airline operating costs.

5.3 SWOT Analysis

This section describes the SWOT analysis of the value-added service potentially provided by the aircraft manufacturer in order to gain competitive advantage on the market. The analysis is shown in Table 2.

Table 2 SWOT analysis

	Strengths	Weaknesses
Internal	CI calculation would be fully customised to gain the best performance/efficiency of the particular aircraft type The manufacturer would take advantage from thorough understanding of the technical aspects of the particular aircraft type and connected costs The service would consider also external factors, e. g. ATC restrictions, airport curfews and weather conditions The service would be offered either for the marginal additional price connected with the order of aircraft, or would be included in the aircraft order price	The service represents additional costs to the aircraft manufacturer The service would be dedicated only to the manufacturer's own aircraft, i. e. if an airline has aircraft also from other manufacturers in its fleet, these would not be included in the service. DOC_T are very individual for each airline — aircraft manufacturer would have to run additional consultations in order to tailor its services for each customer The service is likely to be uneconomic when offered free of charge also in case of small purchase orders Thorough analysis of the potential demand for such service and its impact on sales has not been performed yet Cost-benefit analysis for such service has not been performed yet
	Opportunities	Threats
External	CI is widely recognised as the best available method for calculating and streamlining DOC in relation to the cost of fuel By offering such services with marginal or no additional cost for the customers, the aircraft manufacturer would gain a competitive advantage on the market There is a potential for large purchase orders due to higher impact of CI methodology when used by larger fleets Airlines are in favour of any cost saving activities in current geo-economic climate	Sophisticated solutions are already available on the market Main world aircraft manufacturers are already active in the field of CI calculation and are able to quickly adopt similar approach thanks to their sufficient financial as well as expert capabilities Competitive offerings are well known among the airlines and have a good reputation, especially the major players: Lido/flight, AirSavings and Pacelab.

As indicated by the SWOT analysis, the crucial element of the value-added services would be to define the product both in terms of the quality of the service and associated costs (i. e. its price). In order to identify the right approach to the market additional analyses have to be performed, namely a detailed market analysis identifying potential demand and market niches, and the cost-benefit analysis of the proposed service. However, these analyses are out of the scope of this project.

Summary

Although the CI methodology is the airlines' approach to control direct operating costs and fuel cost, there are several issues:

- High costs associated with the implementation and maintaining of the processes;
- Demanding staffing requirements;
- Issues with the proper identification and calculation of DOC_T;
- The lack of thorough knowledge from the side of airlines and their pilots.

There are several dedicated flight planning software currently available on the market from AirSavings, Pacelab and Lido to name a few.

However, all these offerings represent additional costs to the airlines. There is, therefore, an opportunity for the aircraft manufacturers to consider CI at the design stage. While such considerations would lead to more expensive aircraft, the benefit would be lower aircraft operating costs.

6   Future design work and Integration of CI

As the price of fuel has increased in recent years, there is a drive for aircraft manufacturers to design aircraft that are lighter and more fuel efficient. A lighter aircraft that burns less fuel, although more expensive when initially purchased, will result in lower direct operating costs, changing the balance between IOC and DOC. This would enable airlines to have more scope during flight planning as their operational economics will not be as restricted by fuel prices as currently is the case.

This section provides an introductory overview of a number of the more predominant design considerations. This list is not exhaustive, nor does it provide great detail; the intent is to introduce the reader to design considerations that can be utilised to lower direct operating costs. It is envisaged that future DOC studies would focus on specific design aspects in detail.

The elements presented as considerations during the design phase are:

- Materials

- Engine design
- Aerodynamic efficiency
- Aircraft systems
- Alternative fuels

6.1 Materials

Selection of lighter materials with superior or comparable performance, such as carbon composites, can significantly reduce the MTOW, and in doing so, improve fuel efficiency. For example, carbon-carbon brake packs are now typical for large aircraft. Their introduction resulted in significant weight reductions, greater wear life and improved gain performance. Today, new development aircraft, such as the Boeing 787 Dreamliner, have been designed using composite structures for approximately 50% of its construction. Boeing states that the use of composite will help reduce the maintenance costs by reducing structure traditionally prone to fatigue and corrosion. It is predicted that the bias towards composites will continue to increase in line with recent technological advancements (use of composites for prime airframe structure such as spars, fuselage sections and landing gear).

Landing gear in particular is a prime focus of current research and development. When in-flight, the landing gear do not contribute positively to the in-flight aircraft performance and as a result, are considered parasitic mass. Weight reduction is therefore key.

With more advanced materials becoming available, concepts such as wing morphing are becoming possible. Unlike conventional aircraft, morphing is achieved without the use of flaps and other control surfaces. Aeroelastic tailoring[31] is the structural optimisation of composite structures that meet the flight performance requirements. Variables include laminate thickness, laminate orientation and number of plies.

It is rare that individual technological advancements provide a distinct improvement in operating costs; the significant benefits are born out of the optimisation of many parameters working in unison toward a common goal. For example, the disadvantage with regards to the damage detection and reparability of composite materials has been one of the main drivers toward the development of integrated health monitoring systems. Once combined, it is possible to optimise the structure for both design and in-service efficiency. For example, integrated structural health monitoring allows for the potential of real time fatigue optimisation and overload event protection which can offer significant in-service cost benefits while also allowing weight optimisation of the initial structural design. By integrating such system capabilities into the FMS, it could be possible to assess and optimise the aircraft structure on a flight-by-flight basis.

Therefore, in summary, composite material technology is predicted to play an ever increasing role within civil aerospace in a range of applications; however this progression will be interwoven with the advancement of the aircraft systems.

6.2 Engine design

Fuel efficiency is dependent on the type of engine selected and how it is used. Traditionally, there has been limited closed loop system level integration between cockpit inputs, engine FADEC systems and the aircraft FMS. As technology continues to evolve, particularly with regards to the monitoring capability, there will be increased scope at the design stage to truly integrate the engine control within the aircraft systems which in turn will allow operating optimisation. In addition to control integration, due consideration should be given to the following engine parameters which directly impact efficiency:

- By-pass ratios
- Compression ratios
- Engine intercooling systems
- Blade cooling
- Low emission combustion systems
- Specific Fuel Consumption
- FADEC system

Appropriate engine selection is a key contributor to through life cost. The new engines for the 787 Dreamliner and A350 have increased compression ratios, improved fan blade aerodynamics, turbine blade cooling (to allow increase temperatures) and combustion systems. FADEC system developments include heightened sensing for engine health monitoring and icing control. These advantages have contributed towards claimed fuel consumption saving of up to 20%. These improvements have a direct impact on CI.

Current research and development is focusing on engine designs that may reduce the current dominance of high by-pass ratio turbofan engines.

Geared turbofan engine are likely to become commonplace. These are ultra-high bypass ratio engines where the LP shaft attaches to the fan through a reduction gearbox. This enables the bigger fan to run at a lower, more optimum speed while maintaining high compressor and turbine speeds. The results are increased efficiency, lower fuel consumption/emissions and noise.

Counter rotating open rotor engines offer the potential to reduce fuel consumption and CO_2 emissions compared to turbofans with equivalent thrust. However, disadvantages include an increase in noise profile, installation of the open rotor onto the aircraft and control of the blade pitch through the power transmission system.

6.3　Aerodynamic efficiency

Aerodynamic performance is dictated by design. Each new development strives to offer marked improvements over its predecessor; likewise there is a continual push to increase performance of legacy designs by introducing enhancement modifications.

Wingtip design is an example where performance improvements are possible for both new development and legacy applications. New winglets designed for modern aircraft increase the effective aspect ratio by interrupting wingtip vortices and improve the fuel efficiency. One example is a blended winglet which is attached to the wing with a smooth curve rather than a sharp angle and is used to reduce interference drag at the wing/winglet junction.

The Sharklet winglet is a curved junction wingtip device designed by Airbus and is expected to improve fuel consumption by 3.5% over longer sectors and has also found to improve the take-off performance of the A320 Family airframe.

The aerodynamic design of the passive winglets such as those described above will be optimised to give a performance improvement over a particular segment of the mission profile or the best compromise over the entire flight. Since various operators may intend to operate their aircraft with significantly different mission profiles, it may be possible for an aircraft manufacturer to offer a number of different winglet options for a common wing that would best suit the typical mission (or typical CI setting) of the operator.

Active winglets[32] can be designed for more than one phase of flight. A standard winglet may just reduce the wingtip vortices during the take-off and landing phases of flight, whereas the active winglet is able to reduce wingtip vortices during the cruise phase as well as take-off and landing phases. The options are to link the winglet attitude to set primary control surface positions or allow independent, real-time active control. This then opens the potential for assessing and optimising aerodynamic configurations based on a flight-by-flight basis. However, adding a new active system to the aircraft obviously has an impact on weight and complexity which must be traded off against the improvements in aerodynamic efficiency.

Changes to the design of engine nacelles and pylons, landing gear, fuselage and the wings have all improved the aerodynamic profile and reduced fuel efficiency. The Rolls Royce Trent 1000 engine has elliptical leading edge blades to control the air entering the engine. The Airbus A350 new design will offer airlines 25% fuel saving compared to today's current aircraft as a result of a redesigned wing, fuselage and updated engine.

In addition to the primary design considerations, operational factors can also contribute to the aerodynamic efficiency. Parasitic drag resulting from 'dirty' wings can have a tangible impact on operating costs (trade-off between aerodynamics performance/aesthetics and cleaning costs) which has led to the development of 'self-cleaning' protective treatments.

6.4　Aircraft systems

Recent advances have provided more optimised and integrated electrical and mechanical aircraft systems. The bias away from the historically dominant mechanical systems towards more electrical aircraft is primarily the result of fly-by-wire and Integrated Modular Avionics (IMA). The demise of federated architectures and fully mechanical systems offered the ability to integrate aircraft systems with a resulting reduction in system weight, maintenance requirements while gaining reliability and safety improvements.

An all-electric aircraft allows the removal of multiple, independent, mixed technology power generation/distribution and control systems and replaces it with common, integrated networks. To maximise the potential of such a platform requires significant advances in technology. Typically, the evolution toward the ideal evolves in stages, for example:

It is predicted that Electro Mechanical Actuation (EMA) will replace all traditional Hydraulic Actuation (HA) for systems such as flight control and landing gear. Although it is currently common for applications such as engine control, the significant benefits are only realised when embodied at an aircraft architecture level. As EMA technology is not sufficiently mature to support large civil aircraft applications (landing gear in particular), interim solutions are being implemented such as Electro Hydraulic Actuation (EHA). This enables the benefits of an IMA platform philosophy, with the localised performance of hydraulic actuation, however although this permits the aircraft platform progression, the full benefits cannot be realised at aircraft level (heavier, potentially less reliable & safe). In other applications, technological advancement has allowed the introduction of electrical actuation in the form of Electrical

Actuated Braking Systems (EABS). This has introduced significant braking performance benefits as each brake can be controlled independently. For example, it is possible to optimise temperature & gain control which in turn can improve aircraft decel performance and increase brake & tyre life. To achieve this level of control, a heightened level of monitoring is required. With this additional monitoring, comes the possibility of integrating traditionally independent systems into a single integrated system (refer to 7.1). To improve further (higher speeds, immune to electromagnetic interference and offers a weight reduction), fly-by-light technology (fibre optics) are replacing wire cables, with current research assessing fly-by-wireless systems.

Therefore, it can be seen that the advancement in system technology and design will have a significant impact on aircraft operational costs.

6.5 Alternative fuels

Aviation contributes to approximately two per cent of all man-made CO_2 emissions. During the past 40 years, the industry has improved its fuel efficiency and reduced its related CO_2 emissions by around 70 per cent. Recently, even more ambitious environmental targets (carbon neutral growth by 2020 and a 50 per cent net reduction of CO_2 emissions in 2050, compared to 2005 figures) have been declared and rulings such as the EU Emissions Trading System (which came into force in January 2012) have been introduced. As a result, there is now a heightened focus towards carbon footprint reduction. To achieve this, numerous airlines and aircraft manufacturers have successfully completed flight tests with a mix of biomass (rather than fossil fuel) and kerosene jet fuel. Sources tested have included Jatropha, Camelina, Algae and used cooking oil. Biomass has to be sourced in a sustainable way that does not interfere with the production of food or water for humans or livestock. Biomass fuel has the same specification as jet fuel and therefore no changes are needed to the fuel system and engines, the freezing and hot flash point remains unchanged and the main benefit is a reduction in the carbon footprint.

7 case studies

The following case studies provide examples of Stirling projects that directly or indirectly improve aircraft performance and reduce operating costs. These are intended to provide a brief overview of the possibilities that can be realised through the aircraft design process. It is envisaged that further, detailed investigations, within areas that are of direct interest, will be conducted as part of the follow-on phases of work.

7.1 Aircraft integrated ground control

Traditionally, ground control results from a series of pilot inputs direct to the relevant systems in order to decelerate the aircraft and manoeuvre the aircraft to the desired location. The effectiveness and efficiency (not necessarily connected) of aircraft ground control is reliant on the pilot appropriateness of action within the allowances of the flight manual procedures. Consequently, aircraft are frequently operated in an inefficient way (for example: exit overshoots, lengthened turnaround times, accelerated brake and tyre wear etc.).

Stirling is involved in the development of a closed-loop system that optimises aircraft ground control. The system integrates the brake and steering control systems (nose wheel steering and brakes), engine thrust control (thrust reversers) and aerodynamic surfaces (rudder/spoilers) to enhance the ground performance. The control system interprets and assesses the pilot inputs and then commands the systems in order to achieve the most effective and/or efficient way to decelerate and manoeuvre the aircraft while on ground.

The benefits include: heightened prognostic capabilities, reductions in landing gear loads (and as a result landing gear mass savings), shortened turnaround times, extension of the tyre and brake in-service life and improved ground manoeuvrability.

7.2 Health monitoring systems

Health monitoring (Structural (SHM) and non-structural (NSHM)) provides the ability to identify when an issue has occurred (diagnostics), and more importantly, if an issue is imminent (prognostics). Both capabilities can significantly reduce operating costs (overload detection, damage detection, fatigue life optimisation, reduced maintenance time, extended maintenance intervals, reduced operational interrupts etc.)

Stirling currently has a team of engineers involved in the research & development of integrated and retrofitable health monitoring system architectures, encompassing SHM and NSHM. The research & development incorporates new & existing sensor technologies covering both prognostics & diagnostics.

7.3 Reducing turn around time

The ability to turn an aircraft around quickly is highly desirable from operating cost perspective. For long-haul aircraft, the passengers/luggage/cargo etc. embarking/disembarking typically dictates the achievable turn-around time.

For short-haul aircraft, the constraint can be aircraft system related (brake temperature, re-fuel time etc.).

Stirling is currently working on an automatic brake cooling system development that monitors, predicts and manages (via brake fans) the brake temperature to ensure optimised temperature management. Inadequate cooling can lead to hot 'brakes' which can inhibit take-off (incurring fines, additional fees and schedule delays), while over cooling carbon brakes accelerates brake wear (increase in maintenance costs) and reduces brake performance.

7.4 Fuel quality management system

Stirling has developed a software tool that models water within fuel systems (condensation & suspension) taking into consideration the system transfer architecture, metallic & composite structures, mission profiles etc. The tool enables the simulation of specific flights to assess the fuel/water ratio in order to mitigate against safety related events (icing for example), reduce microbial contamination, support inerting system management and also to improve maintenance procedures by defining which tanks require draining/purging/scavenging (the tool will define the state the water is in which will dictate the remedial action) and at what interval.

7.5 Wing morphing (active winglets)

The active winglet is a morphing wing-tip device that manoeuvres around its yaw, pitch and roll axes during all phases of flight. The movement is scheduled independently through the use of a central control unit. By altering their shape, orientation and incidence, the winglets are able to reduce drag during all phases of flight. Stirling is currently involved in a collaborative research project investigating the optimal design of such a system and the practical implications of installing a system on an aircraft.

8 Conclusions and recommendations

In conclusion, this paper has studied how the various costs associated with the operation of an aircraft can be categorised.

The paper in particular discussed Cost Index. It covered how airlines use CI strategically and operationally, and what opportunities there might be for product differentiation by aerospace manufacturers by considering CI more explicitly at the design stage. There is also a comparison of the different flight planning software currently on offer and how it affects CI and operating costs.

In particular there is a discussion on how the FMS uses CI and other parameters to compute the flight profile using lookup tables. The lookup tables are generated based on performance calculations resulting in speed and altitude optimization to minimize DOC.

The aim of this study is to provide an overview of Direct Operating Costs in sufficient detail as to allow aerospace manufacturers to start to identify specific areas for further in-depth research.

It is proposed that future work will involve a more detailed review of the techniques discussed in this paper and how they can be improved and integrated with the design process.

References

[1] Doganis R. , Flying off Course, 3rd edition, Routledge, 2002.

[2] Air Transport Association (ATA), Quarterly Cost Index for Q3 2010, 2010.

[3] Holloway S. , Straight and Level: Practical Airline Economics, 3rd edition, Ashgate, 2008.

[4] Max M. , Leveraging Process Documentation for Time-Driven Activity Based Costing, Journal of Performance Management 20, 2007.

[5] Boeing, Fuel Conservation Strategies: Cost Index Explained, 2010.

[6] Airbus, Getting to Grips with Fuel Economy, Issue 3,2004.

[7] Cook A. & Tanner G. , Modelling the Airline Costs of Delay Propagation, 2011.

[8] Speyer J. , AirSavings: A Perspective to Get Started with Dignified Cost Index Values, presentation, 2006.

[9] http://www. fwz. aero/products/airsavings. html, 7th December 2011.

[10] http://www. iata. org/whatwedo/aircraft_operations/fuel/pages/fuel_conservation. aspx, 7th December 2011.

[11] Lufthansa Consulting, Cost Savings through Operational Excellence in the Environment of Volatile Fuel Prices, presentation, 2009.

[12] IATA, IATA Operations SO&I: Fuel Conservation Projects, commercial presentation, 2007.

[13] IATA, Guidance Material and Best Practices for Fuel and Environmental Management, 1st edition, 2004.

[14] Fokker Services, Fuel and Environmental Management, 2010.

[15] Airbus, Getting to Grips with the Cost Index, Issue 2, 1998.

[16] http://www. pace. de/products/aircraft-performance/pacelab-ci-ops. html,9 December 2011.

[17] Standard Method of Estimating Comparative Direct Operating Costs of Turbine Powered Transport Aircraft, Air Transport Association of America, December 1967.

[18] Short-Medium Range Aircraft AEA Requirements, December 1989.

[19] Operating Costs, http://adg. standford. edu/aa241/cost/cost. html, 6th December 2011.

[20] Helios, Fuel Cost Index Evaluation, Ivan Baruta, 19th January 2012.

[21] Continuous improvement in aircraft fuel efficiency, www. icao. int, 6th December 2011.

[22] 787 Design Highlights, www. newairplane. com/787/, 7th December 2011.

[23] http://www. aircore-systems. com/cms_aircore/index. php? option＝com_content&task＝view&id＝36&Itemid＝45&lang＝english, 12th December 2011.

[24] http://www. sita. aero/product/flight-planning, 12th December 2011.

[25] http://www. lhsystems. com/solutions/airline-solutions/airline-operations-solutions/lido-flightplanning. htm, 13th December 2011.

[26] http://www. lhsystems. com/solutions/airline-solutions/airline-operations-solutions/lido-flight-modules. htm, 13th December 2011.

[27] http://www. aasi. com/products. html, 13th December 2011.

[28] AEA, Operating economy of AEA Airlines, presentation, November 2004.

[29] Peter Horder, SH&E International Air Transport Consultancy, Airline Operating Costs, presentation, January 2003.

[30] Keith Macgregor, Aircraft Operating Economics, presentation, February 2009.

[31] J. E. Herencia, P. M. Weaver, M. I. Friswell, 'Morphing wing design via aeroelastic tailoring', 48th AIAA/ASME/ASCE/AHS/ASC Structures, Structural Dynamics, and Materials Conference.

[32] N. M. Ursachem, A. T. Isikveren, T. Melin, M. I. Friswell, 'Technology Integration for ACTIVE Poly-morphing Winglets, ASME Conference on Smart Materials, Adaptive Structures and Intelligent Systems, 2008.

[33] Airbus, Getting to Grips with Aircraft Performance, Issue 1,2002.

大飞机出版工程
书 目

《民用飞机系统安全性设计与评估技术概论》

《民用航空器噪声合格审定概论》

《机载软件研制流程最佳实践》

《民用飞机金属结构耐久性与损伤容限设计》

《机载软件适航标准 DO‐178B/C 研究》

《运输类飞机合格审定飞行试验指南》(编译)

《民用飞机复合材料结构适航验证概论》

《民用运输类飞机驾驶舱人为因素设计原则》

四期书目

《航空燃气涡轮发动机工作原理及性能》

《航空发动机结构》

《航空发动机结构强度设计》

《风扇压气机气动弹性力学》(英文版)

《燃气轮机涡轮内部复杂流动机理及设计技术》

《先进燃气轮机燃烧室设计研发》

《燃气涡轮发动机的传热和空气系统》

《航空发动机适航性设计技术导论》

《航空发动机控制》

《气动声学基础及其在航空推进系统中的应用》(英文版)

《叶轮机内部流动试验和测量技术》

《航空涡轮风扇发动机试验技术与方法》

《航空轴流风扇压气机气动设计》

《燃气涡轮发动机性能》(译著)

其他书目

《民用飞机环境监视系统》

《民用飞机飞行管理系统》

《飞机内部舒适性设计》(译著)

《航空航天导论》

《航空计算工程》

《涡动力学》(英文版)

《尾涡流控制》(英文版)

《动态工程系统的可靠性分析:快速分析方法和航空航天应用》(英文版)

《国际航空法导论》(译著)

XIZI 西子航空

浙江西子航空工业有限公司
ZHEJIANG XIZI AVIATION INDUSTRY CO., LTD.

浙江西子航空工业有限公司位于浙江省杭州市，公司成立于2010年3月2日，当前注册资本28000万元人民币。公司拥有德国进口五轴数控加工中心、三坐标测量机、激光跟踪仪、钣金液压成型机等设备，主要业务范围为各类飞机零部件的机械加工、钣金加工、热表处理和部件装配。

电话：0571-86096232
邮箱：xiziac@xiziuhc.com
www.xiziac.com

框板 / 连接杆

三坐标测量仪

德国巨浪五轴数控机加中心

4米三轴数控机加中心

沈阳西子航空产业有限公司
SHENYANG XIZI AVIATION INDUSTRY CO., LTD.

沈阳西子航空位于辽宁省沈阳市，成立于2009年9月16日，注册资金6000万元人民币。拥有数控加工中心、净化间、自动下料机、热压罐、C扫描等生产设备和复合材料试验室。主要业务有航空航天复合材料零件制造；波音、空客、庞巴迪等转包金属零件加工；飞机大部件装配等。

电话：024-89794025　　024-89794318

德国肖茨热压罐

机加车间

美国格博下料机

塞斯纳L162机身装配

浙江西子航空紧固件有限公司
ZHEJIANG XIZI AEROSPACE FASTENERS CO., LTD.

浙江西子航空紧固件有限公司隶属于中国500强企业之一的西子联合。公司位于浙江省海宁市，成立于2011年10月20日。主要引进美国、德国等先进航空紧固件制造设备，具有精密数控机加、冷镦、搓丝、不锈钢钝化、热处理、理化试验及检测等能力，主要制造航空新型紧固件和高端标准件。

电话：0573-89235728
邮箱：sales_xafc@xiziuhc.com

全自动表面处理生产线

美国进口5模5冲冷镦机

XA4623　XA3524　XA2663　XA2662
抽芯铆钉

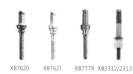
XB7620　XB7621　XB777X　XB2312/2313
抽芯螺栓

安托集团

安托集团成立于1978年。自引进中国第一套CATIA软件至今，已为中国航空制造业信息化建设服务超过20年。我们矢志于为中国航空制造业提供最优秀的产品全生命周期管理解决方案，为民族制造业的进步与腾飞贡献力量！

多年以来，安托伴随着中国航空制造业的发展而成长，目前已经成长为可以为用户提供覆盖产品设计、制造及客服等产品研制全生命周期数字化解决方案的供应商。

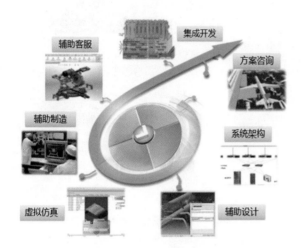

我们的业务领域包括：

1. 方案咨询　　5. 辅助设计
2. 系统架构　　6. 辅助制造
3. 集成开发　　7. 辅助客服
4. 虚拟仿真

设计方法学

任何软件的应用与客户化都首先应当基于用户产品的设计方法学。对于企业来讲，真正适用的方法学一定是将其产品特点、技术理念与软件功能相结合的产物。

- ✓ MBD方法
- ✓ 关联设计方法
- ✓ 构型管理方法
- ✓ 多方案设计方法
- ✓ 并行协同方法
- ......

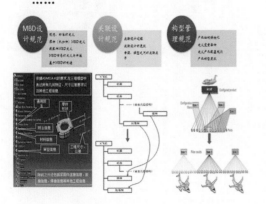

资源库

丰富易用的资源库，是设计效率与质量提升的重要支撑。基于知识工程技术与计算机软件技术，企业可以将众多种类的模型、文档、表格乃至存储于大脑中的经验知识转化为显性可用的设计资源。

- ✓ 统一标准件库
- ✓ 知识工程模板库
- ✓ 标准材料库
- ✓ 工具设备库
- ......

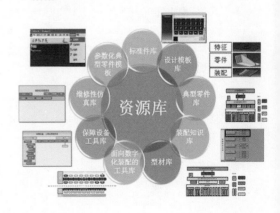

设计辅助

- ✓ 紧固件轻量化设计
- ✓ 飞机外形快速设计
- ✓ 三维标注辅助
- ✓ 模型质量检查
- ✓ 模型变更自动比对
- ✓ 产品更改结构化表达
- ✓ 快速建模辅助
- ✓ 产品色标管理
- ✓ VPM客户功能拓展
- ✓ CATIA复材功能拓展

……

制造辅助

- ✓ 三维CAPP系统
- ✓ 三维工艺设计与仿真
- ✓ 紧固件工艺设计工具
- ✓ 弯管加工辅助工具
- ✓ 工装快速设计软件
- ✓ 快速NC编程软件
- ✓ 制造执行系统MES

……

客服辅助

- ✓ 交互式产品手册制作系统
- ✓ Helpdesk系统
- ✓ 3D课件制作系统

……

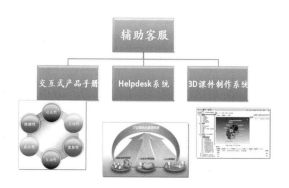

安託集團　ATOZ GROUP

地址：上海市松江松新飞路1500号23号楼
电话：（021) 6760 1028
网址：www.atoz.com.cn

CETC Avionics 中电科航空电子有限公司

公司简介

中电科航空电子有限公司（简称"电科航电"）是中国电子科技集团公司（简称"中国电科"）通过整合内部优质资源，与四川省、成都市、成都市高新区通力合作共同出资组建，专门从事民用航空电子系统及设备业务的高新技术企业。公司注册资本20亿元，于2009年6月16日成立，总部设在成都。

电科航电主要从事民用飞机机载航空电子、通用飞机／无人机系统、空中交通管理、维修保障和训练模拟等业务。公司以国内控股、国际合作的方式，下辖成都华太航空科技有限公司、中电科泰雷兹航空电子有限公司、中电科柯林斯航空电子有限公司等下属公司。

电科航电作为中国电科航空电子产业的发展平台，肩负着国家发展自主民机航电系统与产业的战略使命。电科航电人将以C919航电系统为抓手，推动五大产业平衡发展，着力培育自主创新能力、适航取证能力和市场核竞争力，将公司打造成为国内民机航空电子系统研发和生产的产业基地，成为国内一流、国际知名的航空电子系统和设备的供应商，树立中国电科民用航电品牌，承担更多的企业社会责任，为国家、地方经济发展和社会进步做出贡献。

发展愿景

电科航电公司将以提升国家自主创新能力和建立我国民用航空工业体系为己任，以集团公司"国内卓越，世界一流"宏伟目标为指引，坚持"一二五"发展思路，锐意进取，开拓创新，努力把公司建设成为"国内一流，国际知名"的航空电子供应商和系统集成商。

"一个目标"：到2020年，把电科航电公司打造成为"国内一流，国际知名"的航空电子供应商和系统集成商。

"两步走"：第一步，各项业务全面起步，基本建成现代企业制度和市场运行机制，初步树立中国电科航电品牌；第二步，研发、生产、适航及客服体系建成运行，公司业务融入全球航电产业链，基本达成公司远景目标。

"五大产业"：机载航空电子、通用飞机／无人机系统、空中交通管理、维修保障、训练模拟。

公司荣誉与资质